MAMMALS

of

ONTARIO

Tamara Eder

Lone Pine Publishing

The Publisher: Lone Pine Publishing

10145 – 81 Avenue
Edmonton, AB, Canada
T6E 1W9

1901 Raymond Avenue SW, Suite C
Renton, WA, USA
98055

Website: www.lonepinepublishing.com

National Library of Canada Cataloguing in Publication Data

Eder, Tamara, 1974–
   Mammals of Ontario

   Includes index.
   ISBN 1-55105-321-7

   1. Mammals—Ontario—Identification. I. Title.
QL721.5.O5E33 2002   599'.09'713   C2002-910349-5

Editorial Director: Nancy Foulds
Project Editor: Lee Craig
Illustrations Coordinator: Carol Woo
Technical Review: Jon (Sandy) Dobbyn
Track Terminology: Mark Elbroch
Production Coordinator: Jennifer Fafard
Cover Design: Rod Michalchuk
Layout & Production: Ian Dawe
Map Work: Ian Dawe, Elliot Engley, Lee Craig
Cover Photo: Red Fox, by Wayne Lynch
Track and Print Illustrations: Ian Sheldon
Scanning, Separations & Film: Elite Lithographers Co.

The publisher and author thank Chris C. Fisher and Don Pattie for their previous contributions to the Mammal series.

Photo and Illustration Credits

All photographs are by Terry Parker, except as follows:
Leslie Degner, p. 66; Mark Degner, pp. 78 & 82; Renee DeMartin/West Stock, pp. 96–97; Eyewire, p. 112; David B. Fleetham/Visuals Unlimited, p. 44; Wayne Lynch, p. 86 & pp. 120–21; Brian Wolitski, p. 108.

All illustrations are by Gary Ross, except as follows:
Kindrie Grove, pp. 107, 145 & 201; Ian Sheldon, pp. 42 & 43.

The photographs in this book are reproduced with the generous permission of their copyright holders.

We acknowledge the financial support of the Government of Canada through the Book Publishing Industry Development Program (BPIDP) for our publishing activities.

PC: *08*

Contents

QUICK REFERENCE GUIDE

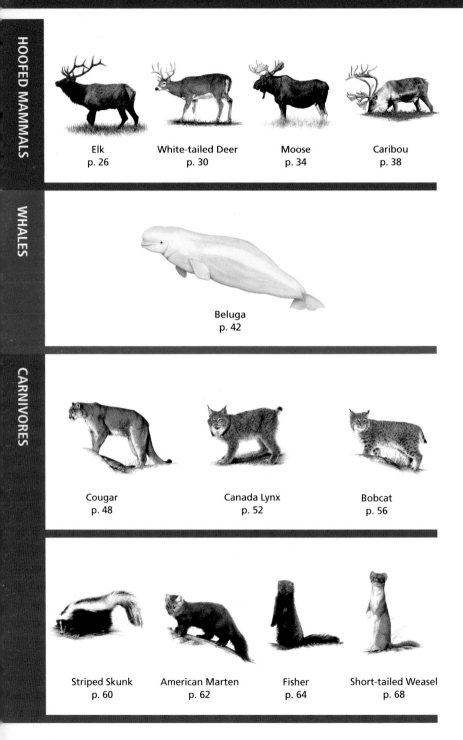

HOOFED MAMMALS

Elk
p. 26

White-tailed Deer
p. 30

Moose
p. 34

Caribou
p. 38

WHALES

Beluga
p. 42

CARNIVORES

Cougar
p. 48

Canada Lynx
p. 52

Bobcat
p. 56

Striped Skunk
p. 60

American Marten
p. 62

Fisher
p. 64

Short-tailed Weasel
p. 68

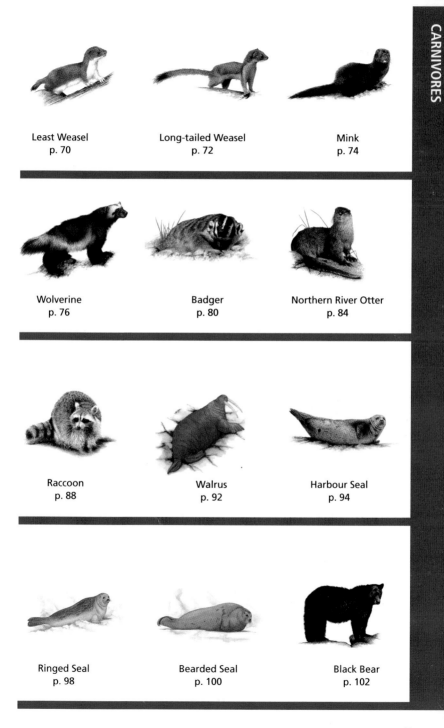

Least Weasel
p. 70

Long-tailed Weasel
p. 72

Mink
p. 74

Wolverine
p. 76

Badger
p. 80

Northern River Otter
p. 84

Raccoon
p. 88

Walrus
p. 92

Harbour Seal
p. 94

Ringed Seal
p. 98

Bearded Seal
p. 100

Black Bear
p. 102

CARNIVORES

Polar Bear
p. 106

Coyote
p. 110

Grey Wolf
p. 114

Arctic Fox
p. 118

Red Fox
p. 122

Grey Fox
p. 126

RODENTS

Porcupine
p. 132

Meadow Jumping
Mouse, p. 136

Woodland Jumping
Mouse, p. 137

Norway Rat
p. 138

House Mouse
p. 140

Deer Mouse
p. 142

White-footed
Mouse, p. 144

Southern Red-
backed Vole, p. 145

Eastern Heather
Vole, p. 146

Meadow Vole
p. 147

Rock Vole
p. 148

Woodland Vole
p. 149

Muskrat
p. 150

Southern Bog
Lemming, p. 152

Northern Bog
Lemming, p. 153

Beaver
p. 154

Eastern Chipmunk
p. 158

Least Chipmunk
p. 160

RODENTS

Woodchuck
p. 162

Franklin's Ground
Squirrel, p. 164

Red Squirrel
p. 166

Eastern Grey
Squirrel, p. 168

Eastern Fox
Squirrel, p. 170

Southern Flying
Squirrel, p. 171

Northern Flying
Squirrel, p. 172

HARES & RABBITS

White-tailed
Jackrabbit, p. 175

European Hare
p. 176

Snowshoe Hare
p. 178

Eastern Cottontail
p. 180

BATS

Northern Bat
p. 183

Little Brown Bat
p. 184

Eastern Small-
footed Bat, p. 186

Eastern Red Bat
p. 187

| Hoary Bat
p. 188 | Silver-haired Bat
p. 190 | Big Brown Bat
p. 191 | Eastern Pipistrelle
p. 192 |

Virginia Opossum
p. 196

Hairy-tailed Mole
p. 198

Eastern Mole
p. 199

Star-nosed Mole
p. 200

Masked Shrew
p. 201

Water Shrew
p. 202

Smoky Shrew
p. 204

Arctic Shrew
p. 205

Pygmy Shrew
p. 206

Northern Short-tailed
Shrew, p. 207

Least Shrew
p. 208

9

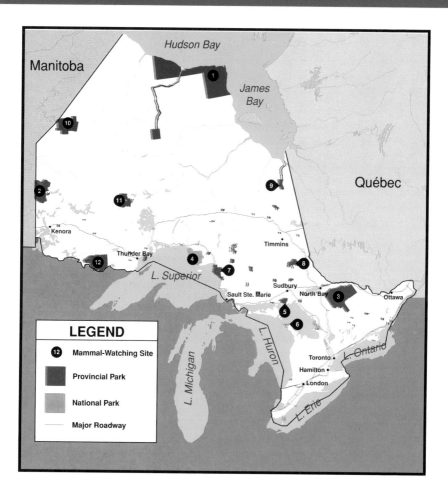

LEGEND

- **12** Mammal-Watching Site
- Provincial Park
- National Park
- Major Roadway

MAMMAL-WATCHING SITES IN ONTARIO

1. Polar Bear Provincial Park
2. Woodland Caribou Provincial Park
3. Algonquin Provincial Park
4. Pukaskwa National Park
5. Killarney Provincial Park
6. Bruce Peninsula National Park
7. Lake Superior Provincial Park
8. Lady Evelyn-Smoothwater Provincial Park
9. Kesagami Provincial Park
10. Opasquia Provincial Park
11. Wabakimi Provincial Park
12. Quetico Provincial Park

Introduction

Few things characterize wilderness as well as wild animals, and few animals are more recognizable than our fellow mammals. In fact, many people use the term "animal" when they really mean "mammal"—they forget that birds, reptiles, amphibians, fish and all the many kinds of invertebrates are animals, too.

Mammals come in a wide variety of colours, shapes and sizes, but they all share two characteristics that distinguish them from the other vertebrates: only mammals have real hair, and only mammals nurse their young from mammary glands (the feature that gives this group its name). Other, less well-known features that are unique to mammals include a muscular diaphragm, which separates the lower abdominal cavity from the cavity that contains the heart and lungs, and a lower jaw that is composed of a single bone on each side. Additionally, a mammal's skull joins with the first vertebra at two points of contact—a bird's or reptile's skull has only one point of contact, which is what allows birds to turn their heads so far around. As well as setting mammals apart from all other kinds of life, these characteristics also identify humans as part of the mammalian group.

Whether you want to watch a Beaver swim in the evening light, catch a glimpse of a Moose in the woods or listen to the haunting sound of a Grey Wolf's howl, Ontario provides many spectacular mammal-watching opportunities. Despite the pressures of human development, much of Ontario's forested interior, Great Lakes region and cold northern realm are internationally recognized destinations for visitors who are interested in rewarding natural experiences.

To honour this treasure is to celebrate North America's intrinsic virtues, and this book is intended to provide readers with the knowledge they need to appreciate the rich variety of mammals in this province. Whether you are a naturalist, a photographer, a wildlife enthusiast or all three, you will find terrific opportunities in Ontario to satisfy your greatest wilderness expectations.

Moose

The Ontario Region

The natural regions of Ontario are extremely diverse. This province, the second largest in Canada, represents a little more than 10 percent of Canada's land area, and it encompasses a dramatic variety of landscapes. Extensive boreal forests, clear blue lakes, major rivers, diverse deciduous forests, expansive tundra and icy coastal waters all contribute to the scenic beauty and ecological uniqueness of this region. Ontario stretches 1685 km from its northernmost point on Hudson Bay to its southernmost point, Pelee Island. Its east-to-west extent is more than 1628 km, giving it an overall area of 1,068,580 km². As much as 75 percent of Ontario is forested, and almost 17 percent is inland water.

The wildlife and wildlife associations that occur in Ontario are linked to the geological, climatic and biological influences of its varied biogeography. For simplification, this book divides the province into five natural regions: Tundra, Taiga, Boreal Forest, Mixed Forest and Carolinian Forest. Looking at these regions in detail can lead to a better understanding of Ontario's mammals and how they interact with each other.

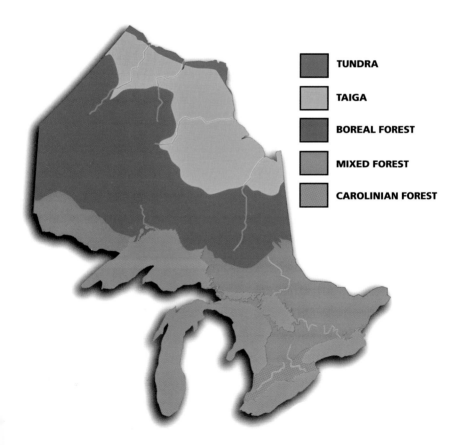

TUNDRA

TAIGA

BOREAL FOREST

MIXED FOREST

CAROLINIAN FOREST

Tundra

The permafrost belt along the Hudson Bay coast in northern Ontario largely defines the Tundra ecoregion. A permafrost zone indicates land that has a subsurface layer of soil that is frozen all year, regardless of snow cover or season. Geologically, the Tundra is part of the Hudson Bay Lowlands, and it is composed primarily of sedimentary rock, such as limestone. The few trees that can grow on the Tundra are small and stunted. Mosses, lichens and low shrubs make up the bulk of the vegetation, but animal life is still abundant. The cold, food-rich water of Hudson Bay supports large numbers of fish, which in turn support large numbers of marine mammals and sea birds. Human habitation in this region is limited to trading villages near the mouths of major rivers. The mammals of this region include the Bearded Seal, Polar Bear, Caribou, Arctic Fox, Grey Wolf, Wolverine and Arctic Shrew.

Caribou

Taiga

The Taiga, also called the Hudson Bay Lowland Forest, is geologically a part of the Canadian Shield (specifically, the Laurentian Lowlands). The underlying igneous rock is mainly granite, and its origin is Precambrian. The result is a flat, poorly drained land that supports extensive swamps, bogs and patterned fens. The typical vegetation is black spruce and tamarack in the muskeg zones. Higher and slightly warmer areas can support some deciduous trees such as balsam poplar, trembling aspen and paper birch. Near the edge of James Bay, black spruce is replaced by white spruce, a tree species better adapted to seashore environments. The Taiga is characterized by long, harsh winters and short, mild summers. Mammals common to the Taiga forest include the Moose, Black Bear, Grey Wolf, American Marten, Short-tailed Weasel, Northern Flying Squirrel, Northern Bog Lemming and Hoary Bat.

Northern Flying Squirrel

13

Boreal Forest

The famous Boreal Forest is the largest natural region in Ontario and the largest forest region in Canada. Geologically, it is also part of the Canadian Shield. The Boreal Forest is very similar to the more northerly Taiga forest, except that its climate is somewhat milder and it is slightly more diverse biologically. Still, snow is present for eight or nine months of the year, and summer temperatures rarely rise above 25° C. The main tree species in the Boreal Forest are black spruce, white spruce, tamarack, jack pine (in well-drained areas) and balsam fir, but some broad-leaved trees, such as paper birch, trembling aspen and balsam poplar, also occur. The pristine lakes and rivers add to the scenic beauty of the Boreal Forest, and tourism here is a year-round interest. This region is also mined for its minerals, including copper, zinc, lead, nickel and uranium. A variety of mammals inhabits the Boreal Forest, such as the Moose, Black Bear, Fisher, Canada Lynx, Snowshoe Hare, Red Squirrel, Beaver, Southern Red-backed Vole and Little Brown Bat.

Beaver

Mixed Forest

South of the Boreal Forest is the Mixed Forest, a transitional region of both coniferous and deciduous trees that is also called the Great Lakes–St. Lawrence Forest. Geologically, this region is at the southern edge of the Canadian Shield and the St. Lawrence Lowlands. The complex vegetation of the Mixed Forest is a mosaic of eastern white pine, red pine, eastern hemlock and eastern white-cedar; some more boreal trees, such as spruce, jack pine, tamarack, balsam fir, trembling aspen and paper birch; and the southern deciduous trees, such as yellow birch, red maple, red oak and American beech. The diversity of vegetation here supports numerous large and small mammals, such as the Grey Wolf, Coyote, Bobcat, Northern River Otter, Long-tailed Weasel, Porcupine, Woodchuck, Woodland Jumping Mouse and Northern Bat.

Grey Wolf

Carolinian Forest

The Carolinian Forest, also called the Eastern Deciduous Forest, is widespread in the eastern United States but is only found in Canada in the southernmost regions of Ontario. The winters here are mild and relatively short, creating a much longer growing season than elsewhere in the province. Geologically, the Carolinian Forest is part of the Great Lakes and St. Lawrence Lowlands, and sedimentary deposits underlie the rich, deep soils typical of this region. The dominant forest trees here are broad-leaved species, such as sugar maple, American beech, American basswood, red ash, white oak and butternut. A few species that thrive here, such as several hickory species, pawpaw, tulip tree, sassafras and Kentucky coffee tree, are found nowhere else in Ontario. Unfortunately, most of the natural forests in this region have been cleared to make way for extensive human settlement, industry and agriculture. The Carolinian Forest is home to many mammals, such as the White-tailed Deer, Grey Fox, Badger, Raccoon, Eastern Grey Squirrel, Southern Flying Squirrel, Eastern Cottontail, Woodland Vole and Virginia Opossum. While a few of these species, such as the Raccoon and White-tailed Deer, have adapted to human habitation, some species are now found only where pockets of the original forest remain.

Badger

Human-Altered Landscapes

The impact of human activity on natural environments is visible throughout Ontario, but it is most noticeable in the south. Cities, roadways, agriculture, forestry and mining are just a few of the impacts we have had on the land. Many of the most common plants and animals that are found in these altered landscapes did not occur here before modern human habitation and transportation. The House Mouse, Norway Rat and European Hare are some of the highly successful exotics that were introduced to North America from Europe and Asia. As well, the distributions of many native mammals have changed as a result of habitat degradation and fragmentation.

Raccoon

Seasonality

The seasons of Ontario greatly influence the lives of mammals. Aside from bats and marine mammals, most species are confined to relatively slow forms of terrestrial travel. As a result, they have limited geographic ranges and must cope in various ways with the changing seasons.

The rising temperatures, melting snow and warm rains of spring bring renewal. Many mammals bear their young at this time of year, and the abundance of food cycles through the food chain: lush new growth provides ample food for herbivores, and the numerous herbivore young become easy prey for the carnivorous mammals. While some small mammals, particularly the shrews and rodents, mature within weeks, offspring of the larger mammals depend on their parents for much longer periods.

During the warmest time of the year, the animals' bodies have recovered from the strain of the previous winter's food scarcity and spring's reproductive efforts, but summer is not a time of relaxation. To prepare for the upcoming winter, some animals must eat large amounts of energy-rich foods to build up fat reserves, while others work furiously to stockpile food caches in safe places.

Least Weasel, winter coat

Autumn is the time when certain bats migrate out of the province. For other mammals, such as the large ungulates, autumn is the time for mating. Some small mammals, however, such as voles and mice, mate every few months or even all year.

Winter differs in intensity and duration between the different regions of the province. In southern areas, winters are mild and not too stressful. In the northern regions, however, winter can be an arduous, life-threatening challenge for many mammals. For herbivores, high-energy foods are difficult to find, often requiring more energy to locate than they provide in return. Their negative energy budget gradually weakens most herbivores through winter, and those not sufficiently fit at the onset of winter end up feeding the equally needy carnivores, which ironically find an ally in winter's severity. Voles and mice also find advantages in the season—an insulating layer of snow buffers their elaborate trails from the worst of winter's cold. Food, shelter and warmth are all found in this thin layer, and the months devoted to food storage now pay off. Winter eventually wanes, and the warmth and life of spring prevails.

An important aspect of seasonality is its effect on the composition of an area's mammal population. You will typically see a different group of mammals in an area in winter than you will in summer. Many mammals, such as ground squirrels and bears, are dormant in winter. Conversely, many ungulates may be more visible in winter because of the lack of foliage and their tendency to enter meadows to find edible vegetation.

Least Weasel, summer coat

Watching Mammals

Many types of mammals are most active at night, so the best times for viewing are during the "wildlife hours" at dawn and at dusk. At these times of day, mammals are out from their daytime hideouts, moving through areas where they can be more easily encountered. During winter, hunger may force certain mammals to be more active during midday. Conversely, when conditions are more favourable during spring and summer, some mammals may become less active during the day.

The protected areas and parks of Ontario offer some of the best opportunities for mammal-watching in the province, but large areas of this province can be considered wilderness, and wildlife is never far. Many of the larger mammals, in particular, can be viewed easily from a vehicle along the many roadways that cut through province. If you walk backcountry trails or hike through the forests, however, you can find yourself right in the homes of certain mammals.

Although people have become more conscious of the need to protect wildlife, the pressures of increased human visitation have nevertheless damaged critical habitats, and some mammals experience frequent harassment. Modern wildlife viewing demands sensitivity and common sense. While some of the mammals that are encountered in Ontario appear easy to approach, it is important to respect your own safety as much as the safety of the animal being viewed. This advice seems obvious for the larger species (although it is ignorantly dismissed in some instances), but it applies equally to small mammals. Honour both the encounter and the animal by demonstrating a respect appropriate to the occasion. Here are some points to remember for ethical wildlife watching in the field:

Little Brown Bat

- Confine your movements to designated trails and roads, wherever provided. Doing so allows animals to adapt to human use in the area and also minimizes your impact on the habitat.
- Avoid dens and resting sites, and never touch or feed wild animals. Baby animals are seldom orphaned or abandoned, and it is against the law to take them away.
- Stress is harmful to wildlife, so never chase or flush animals from cover. Use binoculars and keep a respectful distance, for the animal's sake and often your own.
- Leave the environment, including both flora and fauna, unchanged by your visits. Take home only pictures and memories.
- Pets are a hindrance to wildlife viewing. They may chase, injure or kill other animals, so control your pets or leave them at home.
- Take the time to learn about wildlife and the behaviour and sensitivity of each species.

Fisher

Ontario's Top Mammal-Watching Sites

Although Ontario has only six national parks, it boasts 280 provincial parks that protect 7.1 million hectares of habitat—almost 9 percent of the province. Understandably, providing details on each of these parks is well beyond the scope of this section on Ontario's wildlife viewing areas. A few of the largest of Ontario's parks are described below. The province's other parks are also worth visiting, however, including the larger provincial parks of Lake Superior, Lady Evelyn-Smoothwater, Kesagami, Opasquia, Wabakimi and Quetico. These parks have varying degrees of accessibility, but for hardy adventurers, their exploration is a thrilling and rewarding pursuit.

Pukaskwa National Park

The Pukaskwa wilderness of Lake Superior's north shore—part of the ancient Canadian Shield landscape—is a region of forested hills, rough ridges and rocky-shored lakes. The majority of the park is forested, and the dominant trees include spruce, fir, cedar, birch and aspen. Human activity within the park is a concern, because protecting this unique part of Ontario's natural heritage is essential. In this wilderness is a small, isolated population of Caribou, as well as good numbers of Grey Wolves, Black Bears, Moose, Mink, Canada Lynx and White-tailed Deer. Several important research projects have taken place here that study the interactions between wolves, Moose and Caribou. Other mammals here include Snowshoe Hares, Porcupines, chipmunks, Northern Flying Squirrels and Woodland Jumping Mice.

White-tailed Deer

Bruce Peninsula National Park

The dolomite limestone cliffs on the shoreline of the Bruce Peninsula are a part of the spectacular scenery for which this park is famous. Beyond the cliffs is a landscape of mixed forests and wetlands, prime habitat for many wildlife species. The diversity of habitats in this small national park allows for a great variety of plant and animal life. Surprisingly, Bruce Peninsula is home to 43 species of orchids. Hiking, camping and skiing opportunities abound, and exploring the park provides excellent opportunities for encounters with the local mammal wildlife. Black Bears, White-tailed Deer, Northern River Otters, Raccoons, Striped Skunks and several squirrel species live within the park boundaries. Bruce Peninsula is also home to a good number of Fishers, and visitors to the park are sometimes rewarded with sightings of these elusive mammals.

Striped Skunk

19

Algonquin Provincial Park

The famous Algonquin park owes much of its scenic beauty to the unique transition environment within its boundaries. Southern areas of the park are mainly deciduous forests, while northern areas mark the beginning of the coniferous boreal forest. The resulting mosaic created by the integration of these two zones is a dramatic and diverse biological landscape. The rough topography also influences this mosaic, and throughout the park you will encounter diverse regions, such as maple forest hills, low spruce bogs, lakes, ponds, streams and rocky ridges. Algonquin's excellent accessibility and enormous biological diversity make it one of the finest wildlife-viewing parks in the province. At least 45 mammal species have been recorded here, and Algonquin is the best place in the province to see Moose or listen to the howl of wolves at night. Other mammals, such as Red Foxes, Fishers, Minks, Northern River Otters and Beavers, also live here and are frequently seen by visitors to the park.

Northern River Otter

Killarney Provincial Park

The rocky and imposing landscape of Killarney Provincial Park has long been a source of inspiration for artists, including some that became members of the Group of Seven. The patchwork of rocky terrain and coniferous forest gives the classic impression of dramatic wilderness, but this park is easily accessible and is located just southwest of Sudbury. Canoeing and back-country hiking are the most common activities in the park. Red Foxes, Northern River Otters, Moose, White-tailed Deer, Bobcats, American Martens and Beavers all inhabit the park and are sometimes encountered by visitors.

Bobcat

Polar Bear Provincial Park

This park, the most northerly park in Ontario, is accessible only by aircraft. The effort required to get there, however, is worth the reward, because this park protects over 2 million hectares of low-lying, unspoiled tundra. Peat soils and muskeg are found over the region, and an obvious treeline occurs in the park. North of the treeline, the plant life includes caribou lichen and reindeer and sphagnum moss; south of the treeline there are stunted willows, spruce and tamaracks. Large numbers of Caribou live here, as do Moose, American Martens, Arctic Foxes, Walruses, Beluga whales, Ringed Seals and Bearded Seals. The high densities of Polar Bears in this park in early winter attract many wildlife enthusiasts, researchers and photographers.

Polar Bear

Woodland Caribou Provincial Park

As its name suggests, Woodland Caribou Provincial Park supports a large number of Caribou. This park is situated along the Manitoba–Ontario border, and access is restricted to air, water or rough forest roads. Canoeing is a major attraction in this park, and the connected lakes and rivers offer more than 1600 km of water routes. The mammals found here are those common to the boreal forest. As well, there are a few species present that are typical of the more southwestern habitats. Other than Caribou, there are also Moose, Black Bears, Beavers, Northern River Otters, Muskrats, Minks, Fishers, Wolverines, weasels, Canada Lynx, Red Foxes and Grey Wolves. This park is also home to one of the only well-documented colonies of Franklin's Ground Squirrels in Ontario.

Wolverine

About This Book

This guide describes 78 species of wild and feral mammals that have been reported in Ontario. Domestic farm animals, such as cattle, sheep and llamas, are not described. Although only one whale species and a few seals are known to occur in the waters of Hudson Bay, only those that can be found near or on the Ontario portion of the coast are included. Humans have lived in this region at least since the end of the last Pleistocene glaciation, but the relationship between our species and the natural world is well beyond the scope of this book.

Organization

Biologists divide mammals (class Mammalia) into a number of subgroups, called orders, which form the basis for the organization of this book. Eight mammalian orders have wild representatives in Ontario: even-toed hoofed mammals (Artiodactyla); whales, dolphins and porpoises (Cetacea); carnivores (Carnivora); rodents (Rodentia); hares and rabbits (Lagomorpha); bats (Chiroptera); insectivores (Insectivora); and opossums (Didelphimorphia). In turn, each order is subdivided into families, which group together the more closely related species. For example, within the carnivores, the Wolverine and the Mink, which are both in the weasel family, are more closely related to each other than either is to the Striped Skunk, which is in its own family.

Mammal Names

Although the international zoological community closely monitors the use of scientific names for animals, common names, which change with time, local language and usage, are more difficult to standardize. In the case of birds, the American Ornithologists' Union has been very effective in standardizing the common names used by professionals and recreational naturalists in North America. There is, as yet, no similar organization to oversee and approve the common names of mammals, which can lead to some confusion.

For example, many people apply the name "mole" to pocket gophers, burrowing mammals that leave loose cores of dirt in fields and reminded early settlers of the moles they knew in Europe. To add to the confusion, most people use the name "gopher" to refer not to pocket gophers, but to ground squirrels. If you consider non-mammalian species, it would get even worse. The name "gopher" is used in many parts of North America to denote a species of snake and even a tortoise!

You may think that such confusion is limited to the less charismatic species of animals, but even some of the best-known mammals are victims of human inconsistency. Most people clearly know the identity of the Moose in our province, but this name can cause great confusion for European visitors. The species that we know as the Moose, *Alces alces*, is called "Elk" in Europe ("elk" and *alces* come from the same root), whereas in North America the "Elk" is *Cervus elaphus*, which is called the "Red Deer" in Europe. The blame for this confusion falls on the early European settlers, who misapplied the name "Elk" to populations of *Cervus elaphus*. In an as-yet-unsuccessful attempt to resolve the confusion, many naturalists advocate using the name "Wapiti" for the species in North America.

Despite the lack of an official list of mammal common names, there are some widely accepted standards, such as the "Revised checklist of North American mammals north of Mexico, 1997" (Jones et al. 1997, Occasional Papers, Museum of Texas Tech University, No. 173). *Mammals of Ontario* follows that checklist for the scientific names and for many of the common names, with the exception of when this book deferred to the expertise and knowledge of our reviewer and what he felt were names more appropriate for Ontario. Readers should also know that other sources have attempted to standardize mammal common names on a worldwide basis.

Range Maps

Mapping the range of a species is a problematic endeavour: mammal populations fluctuate, distributions expand and shrink, and dispersing or wandering individuals are occasionally encountered in unexpected areas. The range maps included in this book are intended to show the distribution of breeding/self-sustaining populations in the region, and not the extent of individual specimen records, except in the case of the Elk (p. 26). While this species is native to Ontario, over-harvesting and loss of habitat, two major factors, wiped it out in the late 1800s. Re-introduction efforts have concentrated on the Bancroft, Lake of the Woods, Sudbury and Blind River areas; these areas are indicated by dots on the range map. For all the range maps, full colour intensity on the map indicates a species' presence; paler areas indicate its absence. For certain species accounts, the range of a mammal is shown by question marks, which indicate uncertainty over the species' presence. Other species have such a small range in Ontario that an arrow is used to draw the reader's attention to the area, and in two species triangles are used to show isolated records.

Absent

Present

Similar Species

Before you finalize your decision on the identity of a mammal, check the "Similar Species" section of the account. It briefly describes other mammals that could be mistakenly identified as the species you are considering. By concentrating on the most relevant field marks, the subtle differences between species can often be reduced to easily identifiable characteristics. As you become more experienced at identifying mammals, you might find you can immediately shortlist an animal to a few possible species. The Similar Species section lets you quickly glean the most relevant field marks to distinguish between those species, thereby expediting the identification process.

For the Bat and Insectivore sections, a basic identification key is included in the description of the family. These keys will be helpful in identifying a species, especially if there is a specimen in hand.

Raccoon

Badger

The
MAMMALS

HOOFED MAMMALS

These mammals include the "megaherbivores" of Ontario: they all fall into the largest size class of terrestrial mammals, and they all eat plants exclusively. All of the native hoofed mammals in the region belong to the order Artiodactyla (even-toed hoofed mammals). All even-toed hoofed mammals have either two or four toes on each limb. If there are four toes, the outer two, which are called dewclaws, are always smaller and higher on the leg, touching the ground only in soft mud or snow. Horses, which are not native to North America, belong to the order Perissodactyla (odd-toed hoofed mammals) and have just a single toe on each foot.

Another difference between the two orders of hoofed mammals is in the structure of their ankle bones. The ankle bones of all even-toed hoofed mammals are grooved on both their upper and lower surfaces, which enables these animals to rise from a reclined position with their hindquarters first. This ability means that the large hindleg muscles are available for fight or flight more quickly than in odd-toed hoofed mammals, such as horses, which must rise front first. As well, the even-toed hoofed mammals all have incisors only on the lower jaw. They have a cartilaginous pad at the front of the upper jaw instead of teeth.

Deer Family (Cervidae)

All adult male cervids (and female Caribou) have antlers, which are bony outgrowths of the frontal skull bones and are shed and regrown annually. In males with an adequate diet, subsequent sets of antlers are larger each year. New antlers are soft and tender, and they are covered with "velvet," a layer of skin with short, fine hairs and a network of blood vessels to nourish the growing antlers. The antlers stop growing in late summer, and as the velvet dries up the deer rubs it off. Cervids are also distinguished by the presence of scent glands in pits just in front of the eyes. Their lower canine teeth look like incisors, so there appear to be four pairs of lower incisors.

Elk

Cervus elaphus

The pitched bugle of a bull Elk is as much a symbol of autumn as the first frost, golden leaves and the honk of migrating geese. Unfortunately, the eerie sounds of male Elk are no longer common in Eastern Canada. The extirpation of this large cervid occurred mainly because of human settlement, agricultural expansion and its exploitation from over-hunting. Although the Elk is no longer a significant part of Ontario's fauna, small numbers of it remain.

The Elk that were originally native to Ontario were Eastern Elk (ssp. *canadensis*), but they are believed to have gone extinct by 1850. Efforts to repopulate Elk in our province began as early as 1897. The only reintroductions that were successful, however, occurred in the 1930s and 1940s. Good numbers of Elk survived from that period, and although some remained in the enclosed areas where they were introduced, many escaped to repopulate neighbouring areas from Burwash to the French River Delta. Unfortunately, in 1949, many of Ontario's Elk were destroyed because of a scare of liver fluke infection (a parasite of deer that might be contracted by cattle).

Since 1996, new reintroduction programs have been aimed at repopulating Ontario's Elk. Releases of Elk from Alberta's Elk Island National Park have already occurred near Burwash and on crown land on Lake Huron's north shore, between Blind River and Bruce Mines. Other areas that might see reintroduced Elk before 2005 include Lake of the Woods, the Haliburton Highlands, the Frontenac Axis and the Ottawa Valley.

Elk form breeding harems to a greater degree than other deer, and the bugle of the male is associated with mating in autumn. A bull Elk that is a harem master expends a considerable amount of energy during the autumn rut—his fierce battles with rival bulls and the upkeep of cows in his harem demand more work than time permits—and, if snows come early, he starts winter in a weakened state. Once the rut is over, however, bulls can put on as much as a pound a day if conditions are good. Cows and young Elk, on the other hand, usually see the first frost while they are fat and healthy. This disparity makes sense in evolutionary terms: cows enter winter pregnant with the next generation, whereas, once winter arrives, the older bulls' major contributions are past.

ALSO CALLED: Wapiti.

RANGE: Holarctic in its distribution, the Elk occupies an enormous belt of chiefly upland forests and grasslands. In North America, it occurs from northeastern B.C. to southern Manitoba and south to California, Arizona and New Mexico. The Elk has been reintroduced into Ontario.

Total Length: 2–2.5 m
Shoulder Height: 1.2–1.5 m
Tail Length: 12–18 cm
Weight: 180–500 kg

male

DESCRIPTION: The summer coat is generally golden brown. The winter coat is longer and greyish brown. Year-round, the head, neck and legs are darker brown, and there is a large, yellowish to orangish rump patch bordered by black or dark brown fur. The oval metatarsal glands on the outside of the hocks are outlined by stiff, yellowish hairs. A bull Elk has a dark brown throat mane, and he starts growing antlers in his second year. By his fourth year, the bull's antlers typically bear six points to a side, but there is considerable variation both in the number of points a bull will have and the age when he acquires the full complement of six. A bull rarely has seven or eight points. The antlers are shed by March. New ones begin to grow in late April, becoming mature in August.

HABITAT: Although the Elk prefers open forests and grasslands, it sometimes

> **DID YOU KNOW?**
>
> Elk can easily run at speeds of 45–48 km/h, and the maximum recorded speed is 72 km/h.

hoofprint (walking)

walking trail

ranges into coniferous forests or brush-lands. In Ontario the Elk mainly inhabits open mixed forest and grasslands.

FOOD: Elk are some of the most adaptable grazers. Woody plants and fallen leaves frequently form much of their winter and autumn diet. Sedges and grasses make up 80 to 90 percent of the diet in spring and summer. Salt is a necessary dietary component for all animals that chew their cud, and Elk may travel great distances to find salt-rich soil.

DEN: The Elk does not keep a permanent den, but it often leaves flattened areas of grass or snow where it has bedded down to sleep.

YOUNG: A cow Elk isolates herself from the herd to give birth to a single calf between late May and early June, following an 8½-month gestation. The young stand and nurse within an hour, and within two to four weeks the cow and calf rejoin the herd. The calf is weaned in autumn.

Moose

SIMILAR SPECIES: The **Moose** (p. 34) is darker and taller and has lighter lower hindlegs. The **White-tailed Deer** (p. 30) has a whitish, rather than yellowish, rump patch and is smaller. The **Caribou** (p. 38) has a whitish, not dark brown, neck, and both sexes have antlers. The antlers of a male Caribou are heavier and have more tines.

White-tailed Deer

Odocoileus virginianus

Given the current status of the White-tailed Deer in Ontario, it is hard to imagine that before the arrival of Europeans this graceful animal was only found in small, isolated populations. Historically, this deer was uncommon in Ontario, but with the spread of agricultural development and forest fragmentation, the White-tailed Deer has become quite widespread. In some parts of the province, the White-tailed Deer is regularly seen in cropland and open fields.

The White-tailed Deer is a master at avoiding detection, so in wilderness areas it can be frustratingly difficult to observe. It is very secretive during daylight hours, when it tends to remain concealed in thick shrubs or forest patches. Once the sun begins to set, however, the White-tailed Deer leaves its daytime resting spot, moving gracefully and weaving an intricate path through dense shrubs and over fallen trees, to travel to a foraging site. Despite its mastery of the terrain, this animal is still vulnerable to predators—the deer is aware that danger could be lurking in any shadow, and its nose and ears continually twitch. The major threats are wolves and humans, although fawns and old or sick individuals may also be easy prey for Coyotes (p. 110).

Speed and agility are good defences against most of the White-tail's predators, but all deer are vulnerable to winters in the northern parts of their range. Snow and a scarcity of high-energy food leave the deer with a negative energy budget from the first deep snowfalls in autumn until green vegetation emerges in spring. In spite of their slowed metabolic rates during winter, some deer may starve before spring arrives; these victims of winter provide food for scavengers.

In national parks and protected areas, White-tailed Deer may become habituated to the presence of humans and can sometimes be closely approached. Doing so can be perilous, however, especially around does protecting their young. A person either approaching or touching a fawn can result in a doe abandoning its fawn. Although real danger exists in approaching any wild animal too closely, reports that White-tailed Deer are responsible for far more human fatalities annually than all North American bears misrepresent this animal. While true, these statistics include human fatalities resulting from vehicle collisions with deer. Each year, several hundred thousand deer are involved in accidents on North American roads.

RANGE: From the southern third of Canada, the White-tailed Deer ranges south into the northern quarter of South America. It is largely absent from Nevada, Utah and California. It has been introduced to New Zealand, Finland, Prince Edward Island and Anticosti Island.

Total Length: 1.4–2.1 m
Shoulder Height: 70–115 cm
Tail Length: 21–36 cm
Weight: 30–115 kg

male

ALSO CALLED: Flag-tailed Deer.

DESCRIPTION: The upperparts are generally reddish brown in summer and greyish brown in winter. The belly, throat, chin and underside of the tail are white. There is a narrow, white ring around the eye and a band around the muzzle. A buck starts growing antlers in his second year. The antlers first appear as unbranched "spikehorns"; in later years, generally unbranched tines grow off the main beam. The main beams, when viewed from above, are usually heart-shaped. The metatarsal gland on the outside of the lower hindleg is about 2.5 cm long.

HABITAT: The optimum habitat for a White-tailed Deer is a mixture of open areas and young forests with suitable cover. This deer frequents valleys, stream courses, woodlands, meadows and abandoned farmsteads with tangled shelterbelts. Areas cleared for roads, parking lots, summer homes, logging and mines support much of the vegetation on which the White-tailed Deer thrives.

DID YOU KNOW?

The White-tailed Deer is named for the bright white underside of its tail. A deer raises, or "flags," its tail when it is alarmed. The white flash of the tail communicates danger to nearby deer and provides a guiding signal for following individuals.

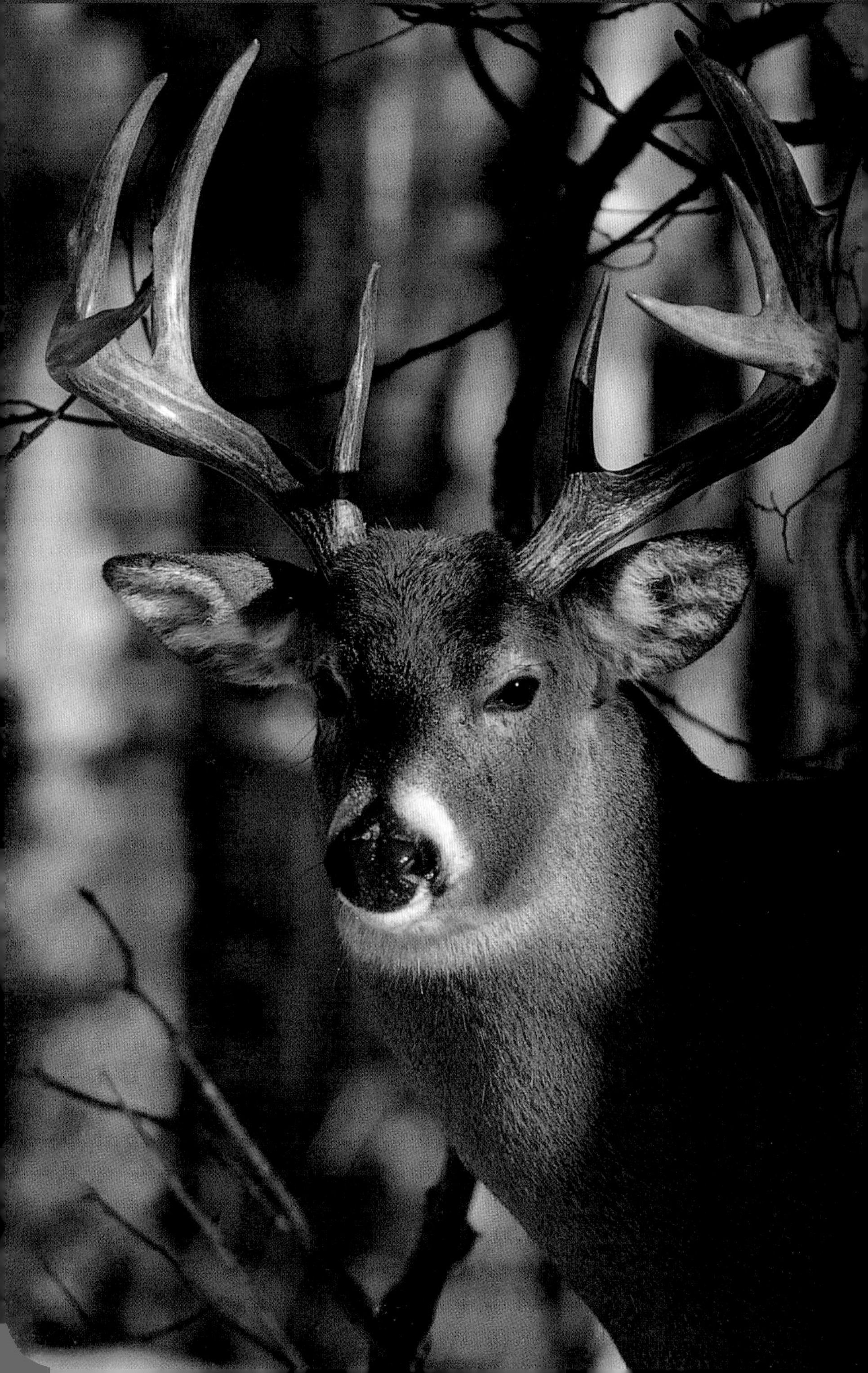

hoofprint (walking)

walking trail

FOOD: During winter, the leaves and twigs of evergreens, deciduous trees and brush make up most of the diet. In early spring and summer, the diet shifts to forbs, some grasses and even mushrooms. On average, a White-tailed Deer eats 2–5 kg of food a day.

DEN: A deer's bed is simply a shallow, oval, body-sized depression in leaves or snow. Favoured bedding areas—often in secluded spots with good all-around visibility where deer can remain safe while they are inactive—will have an accumulation of new and old beds.

YOUNG: A White-tailed doe gives birth to one or two fawns (rarely three) in late May or June, after a gestation of $6\frac{1}{2}$ to 7 months. At birth, a fawn weighs about 2.9 kg, and its coat is reddish with white spots. The fawn can stand and suckle shortly after birth, but it spends most of the first month lying quietly under the cover of vegetation. It is weaned at about four months. A few well-nourished females may mate as autumn fawns, but most do not mate until their second year.

Caribou

SIMILAR SPECIES: The **Caribou** (p. 38) is larger, and its colour is brownish gray to white rather than reddish brown. The male Caribou also has much larger antlers. The **Elk** (p. 26) is larger, and its rump patch is yellowish.

Moose

Alces alces

The monarch of Canadian forests and lush wetlands, the Moose is a handsome animal that provides a thrilling sight for both tourists and wildlife enthusiasts alike. People who know it only from TV cartoon characterizations may not have such feelings for the Moose, but those who have followed its trails through snow and mosquito-ridden bogs respect its abilities.

The Moose is the largest living deer in the world, and Ontario's wildlands offer excellent opportunities to see this majestic ungulate. Recent studies have shown the Moose population in Ontario to be quite dynamic. In general, Moose appear to be increasing in number and range, but in certain southern regions Moose numbers may be low. Factors contributing to low Moose densities in these southern areas include young, open forests, which are more suitable for deer, and parasites.

The Moose's long legs, short neck, humped shoulders and big, bulbous nose may lend the animal an awkward appearance, but these features all serve it well in its environment. With its long legs, the Moose can easily step over downed logs and forest debris and cross streams. Snow, which seriously impedes the progress of other deer and predators, is no obstacle for the Moose, which lifts its legs straight up and down to create very little snow drag.

The Moose has a huge battery of upper and lower cheek teeth, which are perfectly suited for chewing the twigs that make up most of its winter diet. The big bulbous nose and lips hold the twigs in place so the lower incisors can nip them off.

Winter ticks are often a problem for Moose. A single Moose can carry more than 200,000 ticks, and their irritation causes the Moose to rub against trees for relief. With excessive rubbing, a Moose will lose much of its guard hair, resulting in the pale grey "ghost" Moose that are sometimes seen in late winter. Winter deaths are usually the result of blood loss to the ticks, rather than starvation—the twigs, buds and bark of deciduous trees and shrubs that form the bulk of the Moose's winter diet are rarely in short supply. The Moose's common name can also be traced to this feeding habit: the Algonquian called it *moz*, which means "twig eater." The animal's summer diet of aquatic vegetation and other greenery seems quite palatable and varied, but even then, more than half the intake is woody material.

DESCRIPTION: The Moose is the largest living deer in North America. The dark,

RANGE: In North America, this holarctic species ranges through most of Canada and Alaska. Its range has southward extensions through the Rocky and Selkirk mountains, into the northern Midwest states and into New England and the northern Appalachians. The Moose is expanding into the farmlands of North Dakota, South Dakota, Alberta and Saskatchewan, from which it was absent for many decades.

Total Length: 2.5–3 m
Shoulder Height: 1.7–2.1 m
Tail Length: 9–19 cm
Weight: 230–540 kg

male

rich brown to black upperparts fade to lighter, often greyish tones on the lower legs. The head is long and almost horse-like. It has a humped nose, and the upper lip markedly overhangs the lower lip. In winter, a mane of hair as long as 15 cm develops along the spine over the humped shoulders and along the nape of the neck. In summer, the mane is much shorter. Both sexes usually have a large dewlap, or "bell," hanging from the throat. Only bull Moose have antlers. Unlike the antlers of other deer, the Moose's antlers emerge laterally, and many of the tines are palmate, meaning they are merged throughout much of their length, giving the antler a shovel-like appearance. Elk-like antlers are common in young bulls (and they are the only type seen in Eurasian individuals today). A cow Moose has a distinct light patch around the vulva. A calf Moose is brownish to greyish red during its first summer.

HABITAT: Typically associated with the northern coniferous forest, the Moose is most numerous in the early successional stages of willows and poplars. In less-forested foothills and lowlands, it frequents streamside or brushy areas with abundant deciduous woody plants.

DID YOU KNOW?

The Moose is an impressive athlete: individuals have been known to run as fast as 55 km/h, swim continuously for several hours, dive to depths of 6 m and remain submerged for up to a minute.

hoofprint

walking trail

In summer, it may range well up into tundra areas.

FOOD: About 80 percent of the Moose's diet is woody matter, mostly twigs and branches. It prefers deciduous trees and shrubs over conifers. In summer, it also feeds on aquatic vegetation. Sometimes a moose sinks completely below the surface of the water to acquire the succulent aquatics, but these never make up a large part of the diet.

DEN: The Moose makes its daytime bed in a sheltered area, much like other members of the deer family, and it leaves ovals of flattened grass from its weight. Other signs around the bed include tracks, large, oval droppings and browsed vegetation.

YOUNG: In May or June, after a gestation of about eight months, a cow bears one to three (usually two) unspotted calves, each weighing 10–16 kg. To avoid wolves, cows often give birth on islands. The calves begin to follow their mother on her daily routine when they are about two weeks old. A few cows breed in their second year, but most will wait until their third year.

Caribou

SIMILAR SPECIES: With its large size and long head, the Moose resembles a bay or black **Horse** (*Equus caballus*) more than any native mammal. The **Caribou** (p. 38) is smaller and lighter in colour, and the bulls do not have the lateral, palmate antlers of a bull Moose. The **Elk** (p. 26) is shorter and lighter brown in colour and has a yellow rump patch.

Caribou
Rangifer tarandus

The Caribou carves out a living in the deep snows and blackfly fens where most other species of deer do not venture. It appears to do best in areas of expansive wilderness that allow it to undergo seasonal migrations between its summer and winter feeding grounds. This specialist is better adapted to cold climates than other deer—even the Caribou's nose is completely furred.

The Caribou's winter coat has hollow guard hairs up to 10 cm long that top a fine, fleecy, insulating undercoat. These hollow guard hairs provide excellent flotation (as well as insulation) when the animal is swimming across rivers and lakes during its migrations. The Caribou's broad hooves serve it well over rough terrain or to dig through snow to expose edible lichens. The bristle-like hairs that cover a Caribou's feet in winter may help prevent the snow from abrading its skin when the animal digs. This feeding strategy has been one of the Caribou's best-known characteristics for centuries—its name comes from the Micmac name *halibu*, which means "pawer" or "scratcher."

Unlike all other North American cervids, both sexes of the Caribou grow antlers. Not all Caribou shed their antlers at the same time: mature bulls shed their large sweeping racks in December; younger bulls retain their antlers until February; and cows keep theirs until April (within a month they are growing a new set). After losing their antlers, the bulls become subordinate to the still-antlered cows, which are then better equipped to defend desirable feeding sites.

The fragmentation of Caribou populations is of serious concern to resource managers, biologists and naturalists. There are a few places where you can be assured of seeing these threatened animals, but they tend to move throughout their large home range. There is a latitudinal movement; in winter they are more frequently seen in the southern parts of their range. In Ontario, their movements are largely in response to food availability.

The Caribou in Ontario tend to be in smaller bands than the Caribou of the high Arctic. The seasonal movements of these small groups of Caribou hardly compare to the incredible migrations of their more northerly counterparts. There was a time when the Caribou of North America were considered four separate species, but now all the North American Caribou and the Reindeer of Eurasia are classified as one species. Ontario has two populations of Caribou:

RANGE: The North American range of this holarctic animal extends across most of Alaska and northern Canada, from the Arctic Islands south into the boreal forest. It extends south through the Canadian Rockies and Columbia and Selkirk mountains.

Total Length: 1.7–2.4 m
Shoulder Height: 0.9–1.7 m
Tail Length: 13–23 cm
Weight: 90–110 kg

one in the tundra region and the other in the central forested area.

DESCRIPTION: In summer, a Caribou's coat is brown or greyish brown above and lighter below, with white along the lower side of the tail and hoof edges. The winter coat is much lighter, with dark brown or greyish-brown areas on the upper part of the head, the back and the front of the limbs. Both sexes have antlers, but a bull's are much larger. Two tines come off the front of each main antler beam; one lower "brow" tine is palmate near the tip and may be used to push snow to the side as the Caribou feeds. All other tines come off the back of the main beam, an arrangement that is unique to the Caribou.

HABITAT: Caribou are primarily found in either of two habitats in Ontario: forests of spruce, fir and pine or the tundra areas in northern parts of the province.

FOOD: The most important food item for Caribou is fruticose lichens, but grasses, sedges, mosses, forbs, mushrooms and other lichens also contribute to the summer diet. In winter, a Caribou paws at the ground for lichens or eats arboreal lichens within easy reach. As well, winter foods include the buds, leaves and bark of both deciduous and evergreen shrubs. This restless feeder

DID YOU KNOW?

Lichens, the Caribou's favourite winter food, grow very slowly and are frequently restricted to tundra and older spruce and fir forests, but a herd's erratic movements typically prevent it from overgrazing one particular area.

CARIBOU

hoofprint

walking trail

takes only a few mouthfuls before walking ahead, pausing for a few more bites and then walking on again.

DEN: Like other cervids, the Caribou's bed is a simple, shallow, body-sized depression, usually in snow in winter and in leaves or grass in summer. In winter, it usually lies with its body at right angles to the sun on exposed frozen lakes—an arrangement that probably helps it absorb more solar energy. Entire herds will sometimes lie in the same orientation.

YOUNG: Calving occurs in late May or June after a gestation of about 7½ months. The female bears one unspotted calf (rarely twins) weighing about 5 kg. It can follow its mother within hours of birth, and it begins grazing when it is two weeks old. A calf may be weaned after a month, but some continue to nurse into winter. A cow usually first mates when she is 1½ years old; most males do not get a chance to mate until they are at least three to four years old.

White-tailed Deer

SIMILAR SPECIES: The rectangular head and heavy body of the Caribou distinguish it from the other members of the deer family. The **White-tailed Deer** (p. 30) is redder, the **Moose** (p. 34) is larger and darker brown, and the **Elk** (p. 26) has a dark brown neck. Both sexes of Caribou bear antlers, and even calves may bear spikes, a feature that distinguishes them from females or young of other deer species.

41

WHALES

All whales belong to the order Cetacea, and they are distinguished from other mammals by their nearly hairless bodies, paddle-like forelimbs, lack of hindlimbs, fusiform bodies and powerful tail flukes. There are at least 80 species worldwide, classified into two suborders according to whether they have teeth (suborder Odontoceti) or baleen (suborder Mysticeti). The toothed whales are far more numerous and diverse, with some 70 species worldwide. There are only 11 species of baleen whales worldwide, but this group comprises the largest cetaceans.

Hudson Bay has a small but stable population of Beluga whales. Because Belugas are not considered common in this region, a boat trip and accurate data on their whereabouts is required to see these whales reliably. Other whales are usually not sighted close to the Ontario coast, although records do exist. The Minke Whale (*Balaenoptera acutorostrata*), the Narwhal (*Monodon monoceros*) and the Bowhead Whale (*Balaena mysticetus*), for example, are ranked as non-breeding accidental visitors in the southern waters of Hudson Bay.

Beluga and Narwhal Family (Monodontidae)

This small family of whales includes only the Beluga and the Narwhal, polar species that follow the seasonal formation and retreat of ice in the Arctic. These two whales are known to intermingle, and temporary nursery colonies have even formed with members from both species. These whales, which are toothed whales, do not have fused neck vertebrae—an unusual characteristic for cetaceans—and they are able to turn their heads, nod and look around in a manner unlike other whales.

Beluga
Delphinapterus leucas

Total Length: up to 5.5 m; average 4 m
Total Weight: up to 1600 kg; average 820 kg
Birth Length: about 1.5 m
Birth Weight: about 82 kg

In the cold northern waters of the world, the elegant white Beluga lives among the blue-white icebergs. This whale is unmistakable, but passing boaters can overlook it amid the floating chunks of ice. The Beluga is the only cetacean with good numbers in Hudson Bay. Often nicknamed "sea canary" by sailors, this loquacious whale has a vast repertoire of whistles, warbles, squeaks, clucks and squeals that is unlike any other cetacean's in the world.

A Beluga produces these sounds in the air passages in its head; these sounds are probably modified and amplified in the "melon," or forehead. The Beluga seems to have the unique ability to bulge or shrink its melon at will, which may account for the large variety of sounds it produces. The melon of the Beluga, like that of other toothed whales, functions primarily for sound amplification and detection for echolocation. Given the size of the

RANGE: The Beluga has a circumpolar distribution in the subarctic and high Arctic seas. Its movements follow the formation and retreat of seasonal ice.

melon in the Beluga, it probably has one of the most sophisticated sonar systems of any whale. A Beluga can navigate in water that is barely deep enough to cover its body, as well as in open ocean.

Despite research efforts, we do not know the significance of the Beluga's calls. We know that certain sounds are associated with courtship and familial greetings, but whether these whales just chatter noisily, or if they are actually communicating with language remains a mystery. In cases where a juvenile has stranded and must wait six hours for a new high tide—risking sunburn, dehydration and attack by a Polar Bear—the sounds made by the pod members in the water are distinctive. In one documented case where a stranded juvenile survived the tide and returned to the bay, the cacophony of squeaks and trills from its pod were distinctly different from their earlier sounds.

An anomalous but stable population of Belugas survives in the St. Lawrence and Saguenay rivers. This population commonly inhabits the St. Lawrence only as far inland as Québec City, but individuals have been seen as far west as Montréal. It is believed that these Belugas remained there after the retreat of the last ice age. Although the population is stable, the whales here may suffer ill health from the terrible pollutants that contaminate their bodies. When the Belugas of the St. Lawrence die, their bodies must be removed and treated like toxic waste because of the toxins stored in their fat.

ALSO CALLED: White Whale, Sea Canary.

DESCRIPTION: An adult Beluga is unmistakable with its white body. It moults annually, and just prior to

DID YOU KNOW?

A Beluga's neck vertebrae are not fused as in most other whales, so it can move and nod its head around to survey its surroundings. It can even raise its head above the water and look around in the air.

surface, they turn their heads left and right as they survey their surroundings. Because Belugas are gleaming white, their activities underwater are visible provided they are not too deep. When they are moulting, they roll and rub themselves on the bottom. In shallow river deltas where hundreds of Belugas can be seen together, their frolicking and playful behaviour is visible. In many cases, Belugas underwater near a whale-watching boat will turn on their side or even upside-down while swimming so they can look up at the human observers on the deck above them.

GROUP SIZE: Belugas are commonly seen in small groups of 5 to 20 members, but at certain times of the year hundreds or even a few thousand of these whales may gather together at river deltas.

moulting, its old skin appears yellow. After the moult, the new skin is gleaming white. A Beluga does not turn white until it is about five or six years old, and as a juvenile it is dark brown, slate blue or pinkish grey. The Beluga has a small, bulbous head, with a very short but prominent beak. The flippers may curl upward slightly, and so may the tips of the flukes. The trailing edge of the flukes and the dorsal ridge may have a brown tinge. The flukes are notched in the middle and may appear convex in shape. Despite its robust appearance, the Beluga is an extremely flexible creature, and it can twist and turn as it dives underwater. Male Belugas are generally larger than females.

BLOW: The blow of the Beluga is low and steamy and can only be seen on calm, cool days. You may have more luck if you listen for the loud puffing sound instead.

OTHER DISPLAYS: Belugas do not breach, but they frequently spy-hop and lobtail. When they rise above the

FEEDING: The Beluga feeds on a variety of sea creatures, such as squid, fish, mollusks and other invertebrates. Although it may feed near the surface, it frequently dives deep into the ocean, sometimes to a depth of 800 m, where it is presumably feeding on deepwater creatures, such as squid.

YOUNG: When a female is ready to give birth, she swims into deltaic calving waters that are one or two degrees warmer than the open sea. A calf stays with its mother for at least two years, and sometimes as long as four years if the mother does not get pregnant immediately after the calf is weaned. Normally, a female gives birth every two to three years. Peak calving periods are in late summer, mating is in early spring or summer, and gestation is about 14 months. Males do not become sexually mature until they are at least eight years old and females when they are four to seven years old.

SIMILAR SPECIES: No other whales regularly occur in southern Hudson Bay.

CARNIVORES

This group of mammals is aptly named, because, while some members of the order Carnivora are actually omnivorous (and eat a great deal of plant material), most of them prey on other vertebrates. These "meat-eaters" vary greatly in size here in Ontario, from the small Least Weasel to the large and muscular Polar Bear.

Cat Family (Felidae)

Canada Lynx

Excellent and usually solitary hunters, all wildcats have long, curved, sharp, retractile claws. Like dogs, cats walk on their toes—they have five toes on each forefoot and four toes on each hindfoot—and their feet have naked pads and furry soles. As anyone who has a housecat knows, the top of a cat's tongue is rough with spiny, hard, backward-pointing papillae, which are useful to the cat for grooming its fur.

Skunk Family (Mephitidae)

Biologists previously placed skunks in the weasel family, but recent DNA research has led taxonomists to group the North American skunks (together with the stink badgers of Asia) in a separate family. Unlike most weasels, skunks are usually boldly marked, and when threatened they can spray a foul-smelling musk from their specialized anal glands.

Striped Skunk

Weasel Family (Mustelidae)

All weasels are lithe predators with short legs and elongated bodies. They have anal scent glands that produce an unpleasant-smelling musk, but, unlike skunks, they use it to mark territories rather than in defence. Most species have been trapped for their thick, long-lasting fur.

Wolverine

Raccoon Family (Procyonidae)

Raccoons are small to mid-sized omnivores that, like bears (and humans), walk on their heels. They are very good climbers. They are best known for their long, banded, bushy tails and distinctive, black facial masks.

Raccoon

Walrus Family (Odobenidae)

The Walrus—the only member of this family—is an unmistakable creature. Like a hair seal (see below), all of its digits have claws and it has no external ears, but like a member of the eared seal family (Otariidae), the Walrus is able to rotate its hindlimbs under its body to help it move on land. Perhaps the most distinguishing feature of the Walrus is its tusks. Both sexes have tusks, although they tend to be much longer in males.

Walrus

Hair Seal Family (Phocidae)

The hair seals are also known as "true seals," and they are believed to share a common ancestor with the mustelids. These seals have hind-flippers that permanently face backwards; they cannot rotate their hips and hindlegs to support the weight of their bodies. All seals are sometimes referred to as pinnipeds.

Bearded Seal

Bear Family (Ursidae)

The three North American members of this family (two of which occur in Ontario) are the world's largest terrestrial carnivores. All bears are plantigrade—they walk on their heels—and they have powerfully built forelegs and a short tail. Although most bears sleep through the harshest part of winter, they do not truly hibernate—their sleep is not deep and their temperature drops only a couple of degrees.

Black Bear

Dog Family (Canidae)

This family of dogs, wolves, Coyotes and foxes is one of the most widespread terrestrial, non-flying mammalian families. The typically long snout houses a complex series of bones associated with the sense of smell, which plays a major role in finding prey and in communication. Members of this family walk on their toes, and their claws are blunt and non-retractile.

Grey Wolf

Cougar
Puma concolor

A pug-mark in the snow or a heavily clawed tree trunk are two powerful reminders that some places in Ontario are still wild enough for the Cougar. This large cat was once found throughout much of North America, but conflicts with settlers and their stock animals resulted in widespread removal of this great feline. Today, loss of habitat and settlement by people are the most serious issues affecting Cougar populations. The Ontario population is estimated at less than 200 individuals, but accurate data is difficult to obtain. Still, it is one of the most widespread carnivores in the Americas. Its common names reflect this distribution: "puma" is derived from the name used by the Incas of Peru; "cougar" comes from Brazil.

The Cougar is generally a solitary hunter, except when a mother is accompanied by her young. When the young are old enough, they follow their mother and sometimes even help her kill—a process that teaches the young how to hunt for themselves. Although the Cougar is capable of great bursts of speed and giant bounds, it often opts for a less energy-intensive hunting strategy. Silent and nearly motionless, a cat will wait in ambush in a tree or on a ledge until prey approaches. By leaping onto the shoulders of its prey and biting deep into the back of the neck while attempting to knock the prey off balance, the Cougar can take down an adult deer or a small Moose. This big cat needs the equivalent of about one large animal a week to survive, and Cougar densities in the wild tend to correlate with ungulate densities. The Cougar is an adaptable creature that may hunt by day or night.

One of the most charismatic animals of North America, the Cougar is a creature many wildlife lovers hope to see. In Ontario, the Cougar is probably restricted to the northern and northwestern wilderness, where few people are present. This elusive cat is a master of living in the shadows, but if you spend enough time in the wilderness, you may one day see a streak of burnished brown flash through your peripheral vision. If this streak was actually a Cougar, you can count yourself among the extremely lucky. Few people—even biologists—ever get more than a fleeting glimpse of these graceful felines. If you startle one, which is quite improbable—it usually knows of your presence long before you know of its—it will quickly disappear from sight. Only the young may come for a closer look at you. Young Cougars, like most young carnivores, are extremely curious and do not yet realize that humans are best avoided.

RANGE: The Cougar formerly ranged from northern B.C. east to the Atlantic and south to Patagonia. In North America, it is extirpated from most areas except the western mountains and the northern tier of the provinces. A tiny population remains in the Everglades, and there are occasional reports from Maine and New Brunswick.

Total Length: 1.5–2.7 m
Shoulder Height: 65–80 cm
Tail Length: 50–90 cm
Weight: 30–90 kg

ALSO CALLED: Mountain Lion, Puma, Panther.

DESCRIPTION: This handsome feline is the only large, long-tailed native cat in Ontario. Its body is mainly buffy grey to tawny or cinnamon in colour, with pale buff or nearly white undersides. Its body is long and lithe, and its tail is more than half the length of the head and body. The head, ears and muzzle are all rounded. The tip of the tail, sides of the muzzle and backs of the ears are black. Some individuals have prominent facial patterns of black, brown, cinnamon and white.

HABITAT: Cougars are found most frequently in remote, wooded, rocky places, usually near an abundant supply of prey species. In Ontario, they likely inhabit areas away from human activity, probably venturing into brushlands and forested regions in the north and west of the province.

FOOD: In Ontario, Cougars rely mainly on Moose and Caribou. Other prey

DID YOU KNOW?

During an extremely cold winter, a Cougar can starve if the carcasses of its prey freeze solid before it can get more than one meal. This cat's jaws are designed for slicing, and it has trouble chewing frozen meat.

foreprint

walking trail

species probably include Beaver, mice, rabbits and birds. In bitter winters, Cougars feed easily on animals weakened by starvation.

DEN: A cave or crevice between rocks usually serves as a den, but a Cougar may also den under an overhanging bank, beneath the roots of a wind-thrown tree or even inside a hollow tree.

YOUNG: A female Cougar may give birth to a litter of one to six (usually two or three) kittens at any time of the year after a gestation of just over three months. The tan, black-spotted kittens are blind and helpless at birth, but their blue eyes open at two weeks. Their mottled coats help camouflage them when their mother leaves to find food. As the kittens mature, they lose their spots and their eyes turn brown or hazel. They are weaned at about six weeks, by which time they weigh about 3 kg. Young Cougars may stay with their mother for up to two years.

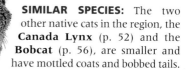

SIMILAR SPECIES: The two other native cats in the region, the **Canada Lynx** (p. 52) and the **Bobcat** (p. 56), are smaller and have mottled coats and bobbed tails.

Canada Lynx

Canada Lynx
Lynx canadensis

Meat is on the nightly menu for the Canada Lynx, and the meal of choice is Snowshoe Hare. The classic predator-prey relationship of these two species is now well known to all students of zoology, but extensive field studies were required to determine how and why these species interact to such a great extent.

Periodic fluctuations in Canada Lynx numbers in local areas have been observed for decades: when hares are abundant, Canada Lynx kittens are more likely to survive and reproduce; when hares are scarce, many kittens starve and the Canada Lynx population declines, sometimes rapidly and usually one to two years after the decline in hares.

The reason why the Canada Lynx is so focused on the Snowshoe Hare (p. 178) as its primary prey may never be understood completely, but the forest community in which this cat lives certainly affects its lifestyle. Many other carnivores compete with the Canada Lynx for the same forest prey. Wolves, Coyotes, Red Foxes, Cougars, Bobcats, Fishers, American Martens, Wolverines, American Minks, owls, eagles and hawks all require animal prey for sustenance, and at least one of these other predators can be found in any habitat where a Canada Lynx lives. Although these other predators may take a hare on occasion, none is as skilled at catching Snowshoe Hares as the Canada Lynx.

This resolute carnivore copes well with the difficult conditions of its wilderness home. Its well-furred feet impart nearly silent movement and serve as snowshoes when snow is deep. Like other cats, the Canada Lynx is not built for fast, long-distance running—it generally ambushes or silently stalks its prey. The ultimate capture of an animal relies on sheer surprise and a sudden overwhelming rush. With its long legs, a lynx can travel rapidly while trailing evasive prey in the tight confines of a forest. It can also climb trees quickly to escape enemies or to find a suitable ambush site.

The Canada Lynx is primarily a solitary hunter of remote forests. During population peaks, however, young cats may disperse into less hospitable environs. In recent memory, the Canada Lynx has been reported within the limits of many major cities. These incidents are unusual, however, because the Canada Lynx typically avoids contact with humans. In Ontario, you are more likely to see the Canada Lynx in forested areas where there is an abundance of hares, but even in this prime territory

RANGE: Primarily an inhabitant of the boreal and mixed forests, the Canada Lynx occurs across much of Canada and Alaska. Its range extends south into the western U.S. mountains and into the northern parts of Wisconsin, Michigan, New York and New England.

Total Length: 80–100 cm
Shoulder Height: 45–60 cm
Tail Length: 9–12 cm
Weight: 7–18 kg

the Canada Lynx is elusive. With each rare observation of a wild lynx, there undoubtedly comes a surprise to people who are accustomed to the appearance of a housecat—the stilt-legged lynx is more than twice the size and gangly in appearance.

DESCRIPTION: This medium-sized, short-tailed, long-legged cat has large feet and protruding ears tipped with 5-cm black hairs. The long, lax, silvery-grey to buffy fur bears faint, darker stripes on the sides and chest and dark spots on the belly and insides of the forelegs. There are black stripes on the forehead and long facial ruff. The entire tip of the stubby tail is black. The long, buffy fur of the hindlegs makes a lynx look like it is wearing baggy trousers. Its large feet spread widely when it is walking, especially in deep snow. The footprint of this cat is big and round, but thick fur often obscures the detail of the print.

HABITAT: The Canada Lynx is closely linked to the northern coniferous forests. Desired habitat components include numerous fallen trees and thickets of young spruce and balsam that serve as effective cover and ambush sites. The Canada Lynx depends on its prey, and its prey depends on the twigs, grasses, leaves, bark and vegetation of the dense forest.

DID YOU KNOW?

Some taxonomists consider the Canada Lynx to be the same species as the European Lynx (*L. lynx*), which occupies the northern forests of Europe and Asia.

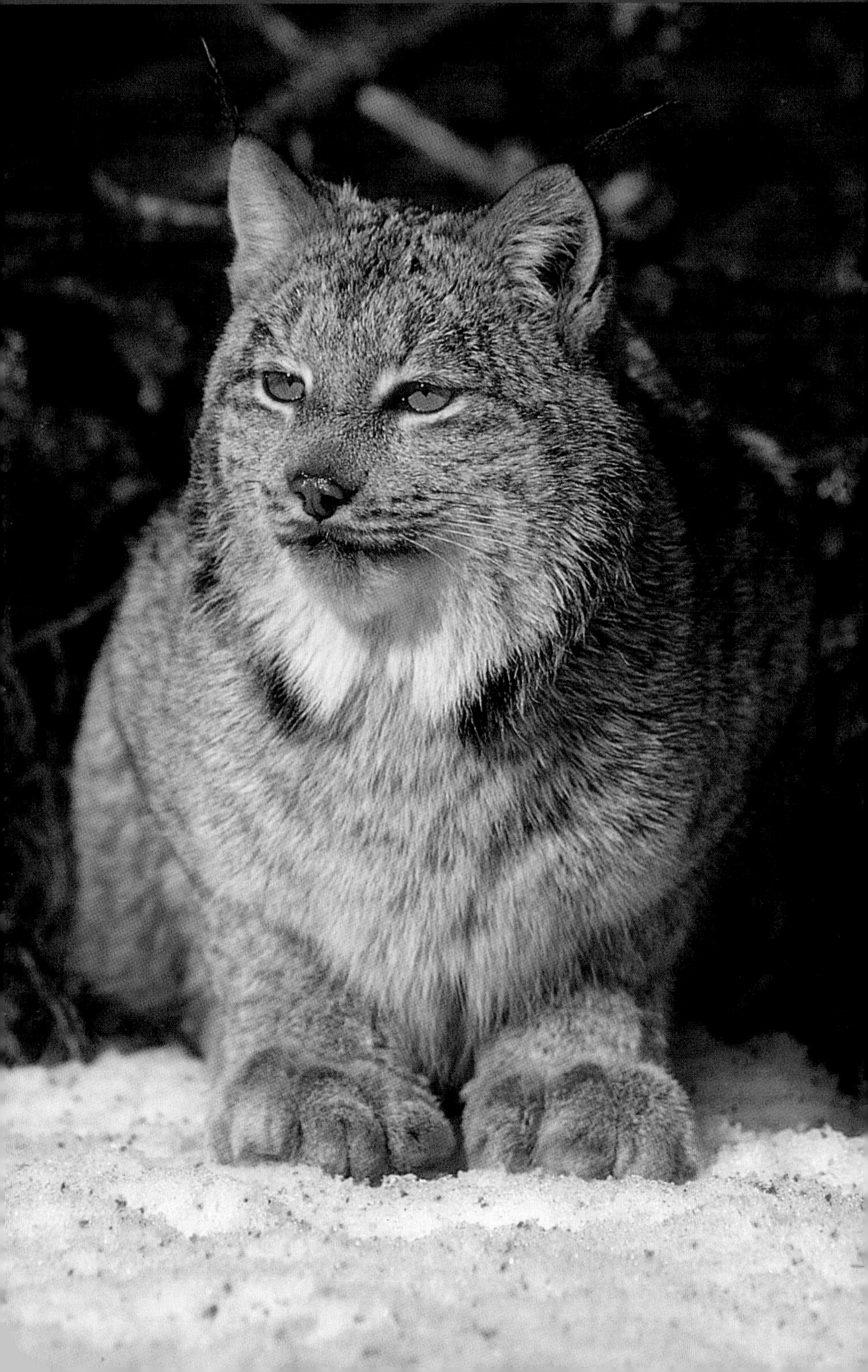

foreprint

walking trail

FOOD: Snowshoe Hares typically make up the bulk of the diet, but a Canada Lynx will sustain itself on squirrels, grouse, rodents or even small domestic animals. When a lynx does not eat all of its kill, it caches the meat under snow or leafy debris.

DEN: The den is typically an unimproved space beneath a fallen log, among rocks or in a cave. Canada Lynx do not share dens, and adult contact is restricted to mating. A mother lynx shares a den with her young until they are mature enough to leave.

YOUNG: Canada Lynx breed in March or April, and the female gives birth to one to five (usually two or three) kittens in May or June. Gestation is a little over two months. The kittens are generally grey, with indistinct longitudinal stripes and dark grey barring on the limbs. Their eyes open in about 12 days, and they are weaned at two months. They stay with their mother through the first winter and acquire their adult coats at 8 to 10 months. A female usually bears her first litter near her first birthday.

Bobcat

SIMILAR SPECIES: The only other native cat that resembles the Canada Lynx is the smaller, shorter-legged **Bobcat** (p. 56). The Bobcat has shorter ear tufts and much smaller feet, and the tip of its tail is black above and white below. The **Cougar** (p. 48) is much larger and has a long tail.

Bobcat
Lynx rufus

For those of us who are naturalists as well as feline enthusiasts, seeing a Bobcat in the wild can be a rewarding experience. The Bobcat is much less common in Ontario than the Canada Lynx (p. 52) because this feline has small feet and cannot cope with the deep snow that covers most of Ontario in the winter. Night drives in parts of northwestern Ontario, such as around Lake of the Woods, offer some of the best opportunities for seeing a Bobcat, although, at best, the experience is a mere glimpse of this cat bobbing along in the headlights.

The Bobcat looks like a large version of a housecat, but it has little of the housecat's domestication. A wildcat in every sense of the word, it impresses observers with its lightfootedness, agility and stealth, usually leaving the momentary experience forever etched in the viewer's mind.

Over the past two centuries, Bobcat populations have fluctuated greatly because of their adaptability to human-wrought change and their vulnerability to our resentment. Less restricted in its prey choices than the Canada Lynx, the Bobcat may vary its diet of hares with any number of small animals, including an occasional turkey or chicken. Its farmyard raids did not go over well with early settlers, and for more than 200 years the Bobcat was considered vermin. Even today, this striking native feline remains on the "varmint" list in some southern states.

Despite its small size, the Bobcat is rumoured to be a ferocious hunter that can take down animals much larger than itself, though some sources still doubt the truth of these stories. This exceptional feat, however, is likely possible: a surreptitious Bobcat waits motionless on a rock or ledge for a deer to approach and leaps onto the neck of the unsuspecting animal. It then maneuvers to the lower side of the neck to deliver a suffocating bite to the deer's throat. Bobcats may resort to such rough tactics when food is scarce, but they usually dine on simpler prey, such as rabbits, birds and rodents. Most of their prey, big or small, is caught at night in ambush. During the day, Bobcats remain immobile in any handy shelter.

Finding Bobcat tracks in soft ground may be the easiest way to determine the presence of this small cat in the region. Unlike Coyote or Red Fox prints, Bobcat prints rarely show any claw marks, and there is one cleft on the front part of the main foot pad and two on the rear. A Bobcat's print is slightly larger than a housecat's, and it tends to be found much farther from human structures.

RANGE: The Bobcat has the widest distribution of any native cat in North America. It occurs in the Rockies and southern interior of B.C., across southern Canada and south through much of the U.S. to Mexico.

Total Length: 75–125 cm
Shoulder Height: 45–55 cm
Tail Length: 13–17 cm
Weight: 7–13 kg

Like all cats, Bobcats bury their scat, and their scratches can help confirm their presence.

DESCRIPTION: The coat is generally tawny or yellowish brown, although it varies slightly with the seasons. The winter coat is usually dull grey with faint patterns. In summer, the coat often has a reddish tinge to it (the source of the scientific name *rufus*). A Bobcat's sides are spotted with dark brown, and there are dark, horizontal stripes on the breast and outsides of the legs. There are two black bars across each cheek and a brown forehead stripe. The ear tufts are less than 2.5 cm long. The chin and throat are whitish, as is the underside of the bobbed tail. The upper surface of the tail is barred, and the tip of the tail is black on top.

HABITAT: The Bobcat occupies open coniferous and deciduous forests and brushy areas, most often along the north shores of Georgian Bay and Lakes Huron and Superior. It especially favours willow stands, which offer excellent cover for its clandestine hunting.

FOOD: The preferred food seems to be hares, but a Bobcat will catch and eat

DID YOU KNOW?

Most cats have long tails, which they lash out to the side to help them corner more rapidly while in pursuit of prey. The Bobcat and the Canada Lynx, however, which typically hunt in brushy areas, have short, or "bobbed," tails that will not get caught in branches.

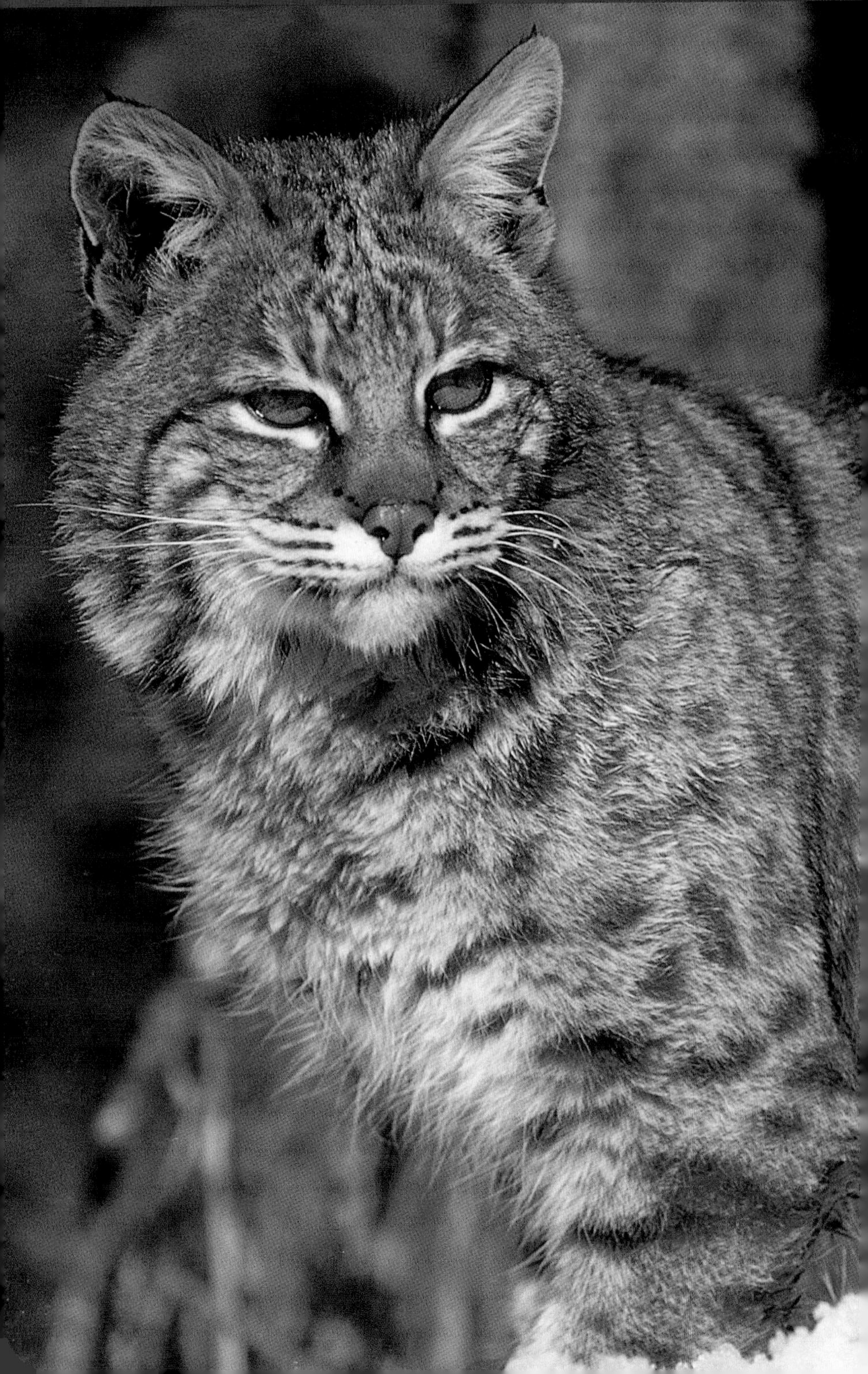

foreprint

walking trail

squirrels, rats, mice, voles, Beavers, skunks, wild turkeys and other ground-nesting birds. When necessary, it scavenges the kills of other animals, and it may even take down its own large prey, such as deer. In areas close to human habitation, it may take domestic cats.

DEN: Bobcats do not keep a permanent den. During the day, they use any available shelter. Female Bobcats prefer rocky crevices for the natal den, but they may also use hollow logs or the cavity under a fallen tree. The mothers do not provide a soft lining in the den for the kittens.

YOUNG: Bobcats typically breed in February or March, giving birth to one to seven (usually three) fuzzy, grey kittens in April or May, but they sometimes breed at other times of the year. The kittens' eyes open after nine days. They are weaned at two months, but they remain with the mother for three to five months. Female Bobcats become sexually mature at one year old, and males at two.

Canada Lynx

SIMILAR SPECIES: The **Canada Lynx** (p. 52) is the only other native bob-tailed cat in North America. These two cats are nearly the same size, but the length of their hindlegs is very different, which makes the Canada Lynx appear taller. The Canada Lynx also has much longer ear tufts, and the tip of its tail is entirely black. The **Cougar** (p. 48) is larger and has a long tail.

Striped Skunk
Mephitis mephitis

In nature, many of the best-dressed creatures are poisonous or unsafe in some way. Poisonous snakes, frogs and insects often bear striking colours and patterns that warn intruders of their dangerous character. The same is true for skunks: their bold black-and-white patterns convey a clear warning to stay away. Anyone not heeding this warning will likely receive a face full of foul, repugnant fluid.

Butylmercaptan is responsible for the stink. Seven different sulfide-containing "active ingredients" have been identified in the musk, which not only smells bad, but also irritates the skin and eyes. Prior to spraying, a distressed Striped Skunk will twist its body into a *U* shape with all four feet on the ground, so that both its head and tail face the threat. If a skunk successfully targets an intruder's eyes, there is intense burning, copious tearing and sometimes a short period of blindness. As well, the musk stimulates nausea in humans.

Despite all these good reasons to avoid close contact with the Striped Skunk, the species is surprisingly tolerant of observation from a discreet distance, and watching a skunk can be very rewarding—its gentle movements contrast with the hyperactive behaviour of its weasel cousins. The Striped Skunk's activity begins at sundown, when it emerges from its daytime hiding place. It usually forages among shrubs, but it often enters open areas, where it can be seen with relative ease. The Striped Skunk is an opportunistic predator that feeds on whatever animal matter is available, and it even digs for grubs, worms and other invertebrates. During winter, its activity is much reduced, and skunks spend the coldest periods in communal dens.

The only regular predator of the Striped Skunk is the Great Horned Owl. Lacking a highly developed sense of smell, this owl does not seem to mind the skunk's odour—nor do the few other birds that commonly scavenge road-killed skunks.

DESCRIPTION: This cat-sized, black-and-white skunk is familiar to most people. Its basic colour is glossy black. A narrow white stripe extends up the snout to above the eyes, and two white stripes begin at the nape of the neck, run back on either side of the midline and meet again at the base of the tail. The white bands often continue on the tail, ending in a white tip, but there is much variation in the amount and distribution of the white markings. The foreclaws are long, and they are used for

RANGE: The Striped Skunk is found across most of North America, from Nova Scotia to Florida in the East and from the southwestern N.W.T. to northern Baja California in the West. It is absent from parts of the deserts of southern Nevada, Utah and eastern California.

Total Length: 55–80 cm
Tail Length: 20–35 cm
Weight: 1.9–4.2 kg

digging. A pair of musk glands on either side of the anus discharges the foul-smelling yellowish liquid for which skunks are famous.

HABITAT: In the wild, the Striped Skunk seems to prefer streamside woodlands, groves of hardwood trees, semi-open areas, brushy grasslands and valleys. It also regularly occurs in cultivated areas, around farmsteads and even in the hearts of cities, where it can be an urban nuisance that eats garbage and raids gardens.

FOOD: All skunks are omnivorous. Insects, including bees, grasshoppers, beetles and various larvae, make up the largest portion, about 40 percent, of the spring and summer diet. To get at bees, skunks will scratch at a hive entrance until the bees emerge and then chew up great gobs of mashed bees, thus incurring the bee-keeper's wrath. The rest of the diet is composed of bird eggs and nestlings, amphibians, reptiles, grains, green vegetation and, particularly in autumn, small mammals, fruits and berries. Along roads, carrion is often an important component of a skunk's diet.

DEN: In most instances, the Striped Skunk builds a bulky nest of dried leaves and grasses in an underground burrow or beneath a building. Winter and maternal dens are underground.

YOUNG: A female Striped Skunk gives birth to 2 to 10 (usually 5 or 6) blind, helpless young in April or May, after a gestation of 62 to 64 days. The typical black-and-white pattern of a skunk is present on the skin at birth. The eyes and ears open at three to four weeks. At five to six weeks, the musk glands are functional. Weaning follows at six to seven weeks. The mother and her young will forage together into the autumn, and they often share a winter den.

SIMILAR SPECIES: The **Badger** (p. 80) has a white stripe running up its snout, but it is larger and squatter and has a grizzled, yellowish-grey body.

DID YOU KNOW?

Fully armed, the Striped Skunk's scent glands contain about 30 ml of noxious, smelly stink. The spray has a maximum range of about 6 m, and a skunk is accurate for half that distance.

American Marten

Martes americana

Ferocity and playfulness are perfectly blended in the American Marten. This quick, active, agile weasel is equally at home on the forest floor or among branches and tree trunks. Its fluid motions and attractive appearance contrast with its swift and deadly hunting tactics. A keen predator, the American Marten sniffs out voles, takes bird eggs, nabs fledglings and acrobatically pursues Red Squirrels.

Unfortunately, this animal's playfulness, agility and insatiable curiosity are not easily observed because it tends to inhabit wilder areas. The American Marten has been known to occupy human structures for short periods of time, should a food source be near, but more typical American Marten sightings are restricted to flashes across roadways or trails. Pursuit of an American Marten rarely leads to a satisfying encounter—this weasel's mastery of the forest is ably demonstrated in its elusiveness.

A close relative of the Eurasian Sable (*M. zibellina*), the American Marten is widely known for its soft, lustrous fur, and it is still targeted on traplines in remote wilderness areas. As with so many species of forest mammals, populations seem to fluctuate markedly every few years—a cyclical pattern revealed by trappers' records. Some scientists attribute these cycles to changes in prey abundance and a corresponding population decline, whereas some trappers suggest that American Marten populations simply migrate from one area to another.

The American Marten is often used as an indicator of environmental conditions because it depends on food found in mature coniferous forests. The loss of such forests has led to declining populations and even extirpation from parts of the province where it was once numerous. Hopefully, modern methods of forest management will maintain adequate habitat for the American Marten and prevent its decline.

ALSO CALLED: Pine Marten.

DESCRIPTION: This slender-bodied, fox-faced weasel has a beautiful pale yellow to dark brown coat and a long, bushy tail. The feet are well-furred and equipped with strong, non-retractile claws. The conspicuous ears are 3.5–4.5 cm long. The eyes are dark and almond-shaped. The breast spot, when present, is usually orange but sometimes whitish or cream, and it varies in size from a small dot to a large patch that occupies the entire region from the chin to the belly. A male is about 15 percent larger

RANGE: The range of the American Marten coincides almost exactly with the distribution of boreal and mixed coniferous forests across North America. It may repopulate where mature forests have developed in areas that were formerly cut or burned.

Total Length: 50–68 cm
Tail Length: 18–23 cm
Weight: 0.5–1.2 kg

than a female. There is a scent gland on the centre of the abdomen.

HABITAT: The American Marten prefers mature, particularly coniferous, forests that contain numerous dead trunks, branches and leaves to provide cover for its rodent prey. It does not occupy recently burned or cut areas.

FOOD: Although rodents such as squirrels, mice and voles make up most of the diet, the American Marten is an opportunistic feeder that will eat hares, bird eggs and chicks, insects, fish, snakes, frogs, carrion and occasionally berries and other vegetation. This active predator hunts both day and night.

DEN: The preferred den site is a hollow tree or log that is lined with dry grass and leaves. American Martens are small enough to refurbish and occupy abandoned woodpecker holes. Females bear their litters in a nest either in a hollow tree, on the ground under protective cover, or in an underground burrow.

YOUNG: Breeding occurs in July or August, but with delayed implantation of the embryo, the litter of one to six (usually three or four) young is not born until March or April. The young are blind and almost naked at birth and weigh just 28 g. The eyes open at six to seven weeks, at which time the young are weaned from a diet of milk to one of mostly meat. The mother must quickly teach her young to hunt, because when they are only about three months old she will re-enter estrus, and, with mating activity, the family group disbands. Young female martens have their first litter at about the time of their second or third birthdays.

SIMILAR SPECIES: The **Fisher** (p. 64) is twice as long, with a long, black tail and often frosted or grizzled-grey to black fur. It seldom has an orange chest patch. The **Mink** (p. 74) has a white chin and irregular white spots on the chest, but it has shorter ears, shorter legs (it does not climb well) and a smaller—and much less bushy—cylindrical tail.

DID YOU KNOW?

Although the American Marten, like most weasels, is primarily carnivorous, it has been known to consume an entire apple pie left cooling outside a window and to steal doughnuts off a picnic table.

Fisher

Martes pennanti

For the lucky naturalist, meeting a wild Fisher is a once-in-a-lifetime opportunity. The rest of us must content ourselves with the knowledge that this reclusive animal remains a top predator in coniferous wildlands. Historically, the Fisher was more numerous, and it once ranged throughout the northern boreal forest, the northeastern hardwood forests and the forests of the Rocky Mountains and the Pacific ranges. In Ontario its numbers are very good, but chance sightings are still limited to secluded areas.

The Fisher is an animal of intact wilderness, and it often disappears shortly after development begins within its range. Forest clearing, habitat destruction, fires and over-trapping resulted in its decline or extirpation over much of its range, but there have been a few reintroductions and a gradual recovery in some areas over the past two decades.

The Fisher is among the most formidable of predators here in Ontario. It is particularly nimble in trees, and the anatomy of its ankles is such that it can rotate its feet and descend trees headfirst. Making full use of its athleticism during foraging, the Fisher incorporates any type of ecological community into its extensive home range, which can reach 120 km across. According to

Ernest Thompson Seton, a legendary naturalist of the 19th century, as fast as a squirrel can run through the treetops, a marten can catch and kill it, and as fast as the marten can run, a Fisher can catch and kill it.

The Fisher is a good swimmer, but, despite its name, it rarely eats fish. Perhaps this misnomer arose because of confusion with the similar-looking—though much smaller—Mink (p. 74), which does regularly feed on fish. These two weasels are most quickly distinguished by their preferred habitats: the mink inhabits riparian areas; Fishers prefer deep forests. In winter, both species can be found in marshy areas, so size and general appearance become the best keys for identification.

Few of the animals on which the Fisher preys can be considered easy picking; the most notable example is the Fisher's famed ability to hunt the Porcupine (p. 132). What the Porcupine lacks in mobility, it more than makes up for in defensive armory, and a successful attack requires all the Fisher's speed, strength and agility. This hunting skill is far less common than wilderness tales suggest, however, and Fishers do not exclusively track Porcupines; rather, they opportunistically hunt whatever crosses their trails. Most of a Fisher's

RANGE: Fishers occur across the southern half of Canada (except the Prairies) and into the northeastern U.S. In the West they are found through the Cascades and Sierra Nevada and from the Rockies to Wyoming.

Total Length: 80–120 cm
Tail Length: 30–40 cm
Weight: 2–5.5 kg

diet consists of rodents, rabbits, grouse and other small animals.

DESCRIPTION: The Fisher has a face that is fox-like, with rounded ears that are more noticeable than those on other large weasels. In profile, its snout appears distinctly pointed. The tail is dark and more than half as long as the body. The coloration over its back is variable, ranging from frosted grey or gold to black. The undersides, tail and legs are dark brown. There may be a white chest spot. A male has a longer, coarser coat than a female, and he is typically 20 percent larger.

HABITAT: The preferred habitats include dense coniferous forests and mixed forests. In winter, Fishers may inhabit marshy areas. They are not found in young forests or where logging or fire has thinned or removed the trees. They are most active at night and thus are seldom seen. Fishers have

extensive home ranges, and they may only visit a particular part of their range once every two to three weeks.

FOOD: Like other members of the weasel family, the Fisher is an opportunistic hunter, killing squirrels, hares, mice, Muskrats, grouse and other birds. Unlike almost any other carnivore, however, the Fisher may hunt the Porcupine, which it kills by repeatedly attacking the head. It also eats berries and nuts, and carrion can be an important part of its diet.

DID YOU KNOW?

The scientific name *pennanti* honours Englishman Thomas Pennant. In the late 1700s, he predicted the decline of the Bison (*Bos bison*) and postulated that Native Americans entered North America via a Bering Strait land bridge.

foreprint

DEN: Hollow trees and logs, rock crevices, brush piles and cavities beneath boulders all serve as den sites. Most dens are only temporary lodging, because the Fisher is always on the move throughout its territory. The natal den is more permanent, and it is usually located in a safe place, such as a hollow tree. A Fisher may excavate its winter den in the snow.

YOUNG: A litter of one to four (usually two or three) young is born in March or April. The mother will breed again about a week after the litter is born, but implantation of the embryo is delayed until January of the following year. During mating, the male and female may remain coupled for up to four hours. The helpless young nurse and remain in the den for at least seven weeks, after which time their eyes open. When they are three months old, they begin to hunt with their mother, and by autumn they are independent. The female usually mates when she is two years old.

walking trail

American Marten

SIMILAR SPECIES: The **American Marten** (p. 62) is smaller and lighter in colour, and it usually has a buff or orange chest spot. The **Mink** (p. 74) is smaller and has shorter ears, shorter legs and a cylindrical, much less bushy tail. The Fisher typically has a more grizzled appearance than either the marten or Mink.

67

Short-tailed Weasel

Mustela erminea

In English, the name "weasel" is often used to describe pointy-nosed villains or to characterize scoundrels. Unfortunately, these connotations give weasels a bad reputation. Although weasels have pointed noses, they are neither villainous nor deceitful. Inarguably, however, weasels are efficient hunters with exceptional predatory talents.

The Short-tailed Weasel is common in Ontario, and it may even be the most abundant land carnivore. Despite its numbers, the Short-tailed Weasel is not commonly seen because, like all weasels, it is most active at night and inhabits areas with heavy cover.

When Short-tailed Weasels roam about their ranges, they explore every hole, burrow, hollow log and brush pile for potential prey. In winter, they travel both above and below the snow in their search for prey. Once a likely meal is located, it is seized with a rush, and then the weasel wraps its body around the animal and drives its needle-sharp canines into the back of the skull or neck. If the weasel catches an animal larger than itself, it seizes the prey by the neck and strangles it.

The Short-tailed Weasel's dramatic change between its winter and summer coats led Europeans to give it two different names: an animal wearing the dark summer coat is called a "stoat"; in the white winter pelage it is known as an "ermine." In Ontario, three weasel species alternate between white in winter and brown in summer, so the stoat and ermine labels are best avoided to prevent confusion. Studies have shown that the trigger for the weasel's colour change is actually decreasing daylight rather than the onset of cold temperatures and snow.

DESCRIPTION: This weasel's short summer coat has brown upperparts and creamy white underparts, often suffused with lemon yellow. The feet are snowy white, even in summer, and the last third of the tail is black. The short, oval ears extend noticeably above the elongated head. The eyes are black and beady. The long neck and narrow thorax make it appear as though the forelegs are positioned farther back than on most mammals, giving the weasel a snake-like appearance. Starting in October and November, these animals become completely white, except for the black tail tip. The lower belly and inner hindlegs may retain some lemon yellow highlights. In late March or April, the weasel moults back to its summer coat.

RANGE: In North America, this weasel occurs throughout most of Alaska and Canada and south to northern California and northern New Mexico in the West and northern Iowa and Pennsylvania in the East.

HABITAT: The Short-tailed Weasel is most abundant in coniferous or mixed forests and streamside woodlands. In summer, it may often be found in the alpine tundra, where it hunts on rockslides and talus slopes.

FOOD: The diet appears to consist almost entirely of animal prey, including mice, voles, shrews, chipmunks, rabbits, bird eggs and nestlings, insects and even amphibians. These weasels are full of curiosity and fury, making them unrelenting in their pursuit of anything they can overpower. They often eat every part of a mouse except the filled stomach, which may be excised with precision and left on a rock.

DEN: Short-tailed Weasels commonly take over the burrows and nests of small rodents (often eating the original occupant first), including mice, ground squirrels, chipmunks, pocket gophers or lemmings, and modify them for weasel settlement. They line the nest with dried grasses, shredded leaves and the fur and feathers of prey. Sometimes a weasel accumulates the furs of so many animals that the nest grows to a diameter of 20 cm. Some nests are located in hollow logs, under buildings or in an abandoned cabin that has a sizable mouse population.

YOUNG: In April or May, the female gives birth to 4 to 12 (usually 6 to 9) blind, helpless young that weigh just 1.8 g each. Their eyes open at five weeks, and soon thereafter they accompany the adults on hunts. At about this time, a male has typically joined the family. In addition to training the young to hunt, he impregnates the mother and all her young females, which are sexually mature at two to three months. Young males do not mature until the next February or March—a reproductive strategy that reduces interbreeding among littermates.

SIMILAR SPECIES: The **Least Weasel** (p. 70) is generally smaller, and, although there may be a few black hairs at the end of its short tail, the entire tip is not black. The **Long-tailed Weasel** (p. 72) is generally larger and has orangish underparts, generally lighter upperparts and yellowish to brownish feet in summer.

DID YOU KNOW?

These weasels typically mate in late summer, but after little more than a week the embryos stop developing. In early spring, up to eight months later, the embryos implant in the uterus and the young are born about one month later.

Total Length:
20–35 cm

Tail Length:
4–9 cm

Weight:
45–105 g

summer coat

Least Weasel
Mustela nivalis

If mice could talk, they would no doubt say that they live in constant fear of the Least Weasel. As it hunts, the Least Weasel enters and fully explores every hole it encounters. This pint-sized carnivore, which is very rare in Ontario—only a few records exist for central and northern Ontario—is small enough to squeeze into the burrows of mice and voles. Any small, moving object seems to warrant attack.

The Least Weasel is the smallest weasel (in fact, the smallest member of the carnivore order) in the world, but it has a monstrous appetite. On average, it consumes about 1 g of meat an hour, which means that it may eat almost its own weight in food each day. If this weasel finds a group of mice or other small rodents, it is quick enough to kill them all within seconds. Prey that is not consumed immediately is stored to be eaten later.

Least Weasels can be active at any time, but they do most of their roaming at night. As a result, few people ever see these animals in action. Most human encounters with a Least Weasel result from lifting plywood, sheet metal or hay bales. These sightings are understandably brief because the weasel wastes little time finding the nearest escape route, and any hole 2.5 cm across or greater is fair to enter—much to the dismay of its current resident.

Because the Least Weasel changes colour in response to shorter day lengths, an unseasonably early snow in the autumn or an early melt in spring can make a weasel stand out against its environs. In spite of this visual disadvantage, a mis-coloured Least Weasel is still an efficient hunter.

DESCRIPTION: In summer, this small weasel is walnut brown above and white below. The short tail may have a few black hairs at the end, but never an entirely black tip. The ears are short, scarcely extending above the fur. In winter, the entire coat is white, including the furred soles of the feet. Only a few black hairs may remain at the tip of the tail.

HABITAT: In Ontario, the Least Weasel usually does not inhabit dense coniferous forests, preferring open, grassy areas, forest edges or tundra. It some-

winter coat

RANGE: In North America, the Least Weasel's range extends from western Alaska through most of Canada and south to Nebraska and Tennessee. It is absent from the Maritimes, New York and New England.

Total Length: 15–22 cm
Tail Length: 2–4 cm
Weight: 25–75 g

summer coat

times occupies abandoned buildings and rock piles. Prey abundance seems to influence the distribution of Least Weasels more than habitat does.

FOOD: Voles, mice and insects are the usual prey, but amphibians, birds and eggs are taken when they are encountered.

DEN: The burrow and nest of a vole that fell prey to a Least Weasel make a typical den site. The nest is usually lined with rodent fur and fine grass, which may become matted like felt and reach a thickness of 2.5 cm. In winter, frozen, stored mice may be dragged into the nest to thaw prior to consumption.

YOUNG: Unlike many weasels, the Least Weasel does not exhibit delayed implantation of the embryos, and a female may give birth in any month of the year, after a gestation of 35 days. A litter contains 1 to 10 (usually 4 or 5) wrinkled, pink, hairless young. At three

weeks they begin to eat meat. After their eyes open at 26 to 30 days, their mother begins to take them hunting. They disperse at about seven weeks, living solitary existences except for brief mating encounters.

SIMILAR SPECIES: The **Short-tailed Weasel** (p. 68) is generally larger, has a longer tail with an entirely black tip and usually has lemon yellow highlights on the belly. The **Long-tailed Weasel** (p. 72) is larger, has a much longer, black-tipped tail and has orangish underparts, generally lighter upperparts and brown feet in summer.

DID YOU KNOW?

During the autumn moult, white fur first appears on the animal's belly and spreads toward the back. The reverse occurs in spring: the brown coat begins to form on the weasel's back and moves toward its belly.

Long-tailed Weasel
Mustela frenata

On a sunny winter day, there may be no better wildlife experience than to follow the tracks of a Long-tailed Weasel. This curious animal zig-zags as though it can never make up its mind which way to go, and every little thing it crosses seems to offer a momentary distraction. The Long-tailed Weasel seems continuously excited, and this bountiful energy is easily read in its tracks as it leaps, bounds, walks and circles through its range.

Long-tailed Weasels hunt wherever they can find prey: on and beneath the snow, along wetland edges, in burrows and even occasionally in trees. They can overpower smaller prey, such as mice, large insects and snakes, and kill them instantly. Larger prey species, up to the size of a rabbit, are grabbed by the throat and neck and then wrestled to the ground. As the weasel wraps its snake-like body around its prey in an attempt to throw it off balance, it tries to kill the animal with bites to the back of the neck and head.

Unlike the Short-tailed Weasel (p. 68) and the more northerly Least Weasel (p. 70), the Long-tailed Weasel only occurs in North America. With the conversion of native grasslands to farmland, the Long-tailed Weasel has been declining throughout much of its range. Still, in some native pastures that teem with rodents, the Long-tailed Weasel can be found bounding about during the daytime, continually hunting throughout its wakeful hours.

DESCRIPTION: The summer coat is a rich cinnamon brown on the upperparts and usually orangish on the underparts. The feet are brown in summer. The tail is half as long as the body, and the terminal quarter is black. The winter coat is entirely white, except for the black tail tip and sometimes an orangish wash on the belly. As in all weasels, the body is long and slender—the forelegs appear to be positioned well back on the body—and the head is hardly wider than the neck.

winter coat

RANGE: From a northern limit in central B.C. and Alberta, this weasel ranges south through most of the U.S. (except the southwestern deserts) and Mexico into northern South America.

Total Length: 28–42 cm
Tail Length: 12–29 cm
Weight: 85–400 g

summer coat

HABITAT: The Long-tailed Weasel is an animal of open country and open woodlands. It may be found in agricultural areas and on grassy slopes. Sometimes it forages in valleys and open forests.

FOOD: Although the Long-tailed Weasel can successfully subdue larger prey than either of its smaller relatives, voles and mice still make up most of its diet. It also preys on ground squirrels, Red Squirrels (p. 166), rabbits and shrews, and it takes the eggs and young of ground-nesting birds when it encounters them.

DEN: These weasels usually make their nests in the burrow of a small mammal they have eaten. Nest cavities are lined with soft materials such as fur from prey or dried grass. The female makes a maternal den in the same manner, either in a burrow, in a hollow log or under an old tree stump.

YOUNG: Long-tailed Weasels typically mate in mid-summer, but, through delayed implantation of the embryos, the young are not born until April or May. The litter contains four to nine (usually six to eight) blind, helpless young. They are born with sparse, white hair, which becomes a fuzzy coat by one week and a sleek coat in two weeks. At 3½ weeks the young begin to supplement their milk diet with meat; they are weaned when their eyes open, just after seven weeks. By six weeks, there is a pronounced difference in size, with young males weighing about 99 g and females about 78 g. At about this time, a mature male weasel typically joins the group to breed with the mother and the young females as they become sexually mature. The group travels together, and the male and female teach the young to hunt. The group disperses when the young are 2½ to 3 months old.

SIMILAR SPECIES: The **Short-tailed Weasel** (p. 68) is typically smaller, has a relatively shorter tail and has a lemon yellow (not orangish) belly and white feet in summer. The **Least Weasel** (p. 70) is much smaller and may have a few black hairs on the tip of its short tail, but it never has an entirely black tip.

DID YOU KNOW?

Weasel signs are not uncommon if you know what to look for: the tracks typically follow a paired pattern, and the twisted, black, hair-filled droppings, which are about the size of your pinkie finger, are often left atop a rock pile.

Mink

Mustela vison

To many people, watching the fluid undulations of a bounding Mink is more valuable than its much-prized fur. The Mink is a lithe weasel that was once described by naturalist Andy Russell as moving "like a brown silk ribbon." Indeed, like most weasel species, the Mink seems to move with the unpredictable flexibility of a toy Slinky in a child's hands.

Minks are tenacious hunters, following scent trails left by potential prey over all kinds of obstacles and terrain. Almost as aquatic as otters, these opportunistic feeders routinely dive to depths of several metres in pursuit of fish. Their fishing activity tends to coincide with breeding aggregations of fish in spring and autumn, or during winter, when low oxygen levels force fish to congregate in oxygenated areas. It is along watercourses, therefore, where Minks are most frequently observed, and their home ranges often stretch out in linear fashion, following rivers for up to 5 km.

The Mink is active throughout the year, and it is often easiest to follow by trailing its winter tracks in snow. The paired prints left by its bounding gait betray the inquisitive animal's adventures as it comes within sniffing distance of every burrow, hollow log and bush pile. This active predator always seems to be on the hunt; scarcely any feeding opportunities are passed up. Minks may kill more than they can eat, and surplus kills are stored for later eating. A Mink's food caches are often tucked away in its overnight dens, which are typically dug into riverbanks, beneath rock piles or in the home of a permanently evicted Muskrat (p. 150).

DESCRIPTION: The sleek coat is generally dark brown to black, usually with white spots on the chin, the chest and sometimes the belly. The legs are short. The tail is cylindrical and only somewhat bushy. A male is nearly twice as large as a female. The anal scent glands produce a rank, skunk-like odour.

HABITAT: The Mink is almost never found far from water. It is comfortable in a variety of habitats from coniferous or hardwood forests to open grassland, provided that lakes, ponds or rivers are available.

FOOD: Minks are fierce predators of Muskrats, but in their desire for nearly any meat they also take frogs, fish, waterfowl, eggs, mice, voles, rabbits, snakes and even crayfish and other aquatic invertebrates.

RANGE: This wide-ranging weasel occurs across most of Canada and the U.S., except for the high Arctic tundra and the dry southwestern regions.

Total Length: 47–70 cm
Tail Length: 15–20 cm
Weight: 0.6–1.4 kg

2 x 2 loping trail

DEN: The den is usually in a burrow close to water. A Mink may dig its own burrow, but more frequently it takes over a Muskrat or Beaver burrow and lines the nest with grass, feathers and other soft materials.

YOUNG: Minks breed any time between late January and early April, but because the period of delayed implantation varies in length (from one week to 1½ months), the female almost always gives birth in late April or early May. The actual gestation is about one month. There are 2 to 10 (usually 4 or 5) helpless, blind, pink, wrinkled young in a litter. Their eyes open at 24 to 31 days, and weaning begins at five weeks. The mother teaches the young to hunt for two to three weeks, after which they fend for themselves.

SIMILAR SPECIES: The **American Marten** (p. 62) has a bushier tail, longer legs and an orange or buff throat patch, and it is not as sleek looking. The **Fisher** (p. 64) is much larger with longer ears and a much bushier tail. The **Northern River Otter** (p. 84) is larger and has a tapered tail and webbed feet.

DID YOU KNOW?

"Mink" is from a Swedish word that means "stinky animal." Although not as aromatic as skunks, Minks are among the smelliest of the weasels. The anal musk glands can release a stinky liquid, but not aim the spray, when a Mink is threatened.

Wolverine

Gulo gulo

The Wolverine is one of the most poorly understood mammals in North America. It is an elusive animal and historically was a creature of many myths and tall tales. More recently, the Wolverine has become a symbol of deep, pristine wilderness. Although most of us will never see a Wolverine, the knowledge that it maintains a hold in remote forests may reassure us that expanses of wilderness still exist.

Tales of the Wolverine's gluttony—its reputation rivals that of hyenas in Africa—have lingered in forest lore for centuries. Pioneers warned their children against the dangers of the forests, and often they meant Wolverines. The Wolverine is an efficient and agile predator: it can crush through bone in a single bite; it has long, semi-retractile foreclaws that allow it to climb trees; and it is ferocious enough to challenge a lone bear or wolf. What we rarely hear about is this animal's intelligence, its uniqueness among its weasel relatives and its sheer vigour and beauty.

From a few behavioral studies of Wolverines, these animals are demystified; instead of vicious gluttons, Wolverines become clever and adaptable creatures. Even simple observations of a Wolverine standing on its hindlegs and scanning the surroundings with a paw at its forehead to shield its eyes from the sun are indicative of intelligent behaviour we are only now starting to understand. Nevertheless, some of the Wolverine's reputation is well deserved. True to its nickname "skunk bear," the Wolverine produces a stink that rivals skunks in foulness. The abundant scent is produced in anal glands and primarily used to mark territory.

The Wolverine's habitat preferences seem to vary as its diet shifts with seasons. In summer, it eats mostly rodents and other small mammals, birds and berries; in winter, it lives on carrion, mainly hoofed mammals, most of which it scavenges from wolves, winter-kills or roadkills. Like a vulture, the Wolverine can detect carcasses from far away.

The largest weasel of all, the Wolverine has one of the mammal world's most powerful set of jaws, which it uses to tear meat off frozen carcasses or to crunch through bone to get at the rich, nourishing marrow inside. Few other large animals are able to extract as much nourishment from a single carcass.

DESCRIPTION: Although the head is small and weasel-like, the long legs and long fur look like they belong on a small bear. Unlike a bear, however, the

RANGE: In North America, the Wolverine is a species of the boreal forest and tundra of Alaska and northern Canada. It follows the montane coniferous forests from Alaska to as far south as California and Colorado.

Total Length: 70–110 cm
Tail Length: 17–25 cm
Weight: 7–16 kg

Wolverine has an arched back and a long, bushy tail. The coat is mostly a shiny, dark cinnamon brown to nearly black. There may be yellowish-white spots on the throat and chest. A buffy or pale brownish stripe runs down each side from the shoulder to the flank, where it becomes wider. These stripes meet just before the base of the tail, leaving a dark saddle.

HABITAT: The Wolverine predominantly occupies large areas of remote, wooded wilderness. In Ontario, they also inhabit the tundra region. The Wolverine's enormous territory encompasses a great variety of habitats; these agile, determined predators are likely able to conquer almost any wild terrain.

FOOD: Wolverines prey on mice, other small mammals, birds, Beavers and fish.

Caribou, even Moose, have been attacked, often successfully. In winter, Wolverines will scavenge malnourished animals or the remains left by other predators. To a limited extent, they eat berries, fungi and nuts. Although Wolverines are generally thought to avoid human habitations, they are known to break into wilderness cabins and meat caches to eat or up-turn everything within.

DID YOU KNOW?

The Wolverine's lower jaw is more tightly bound to its skull than most other mammals' jaws. The articulating hinge that connects the upper and lower jaws is wrapped by bone in adult Wolverines, and for the jaws to dislocate this bone would have to break.

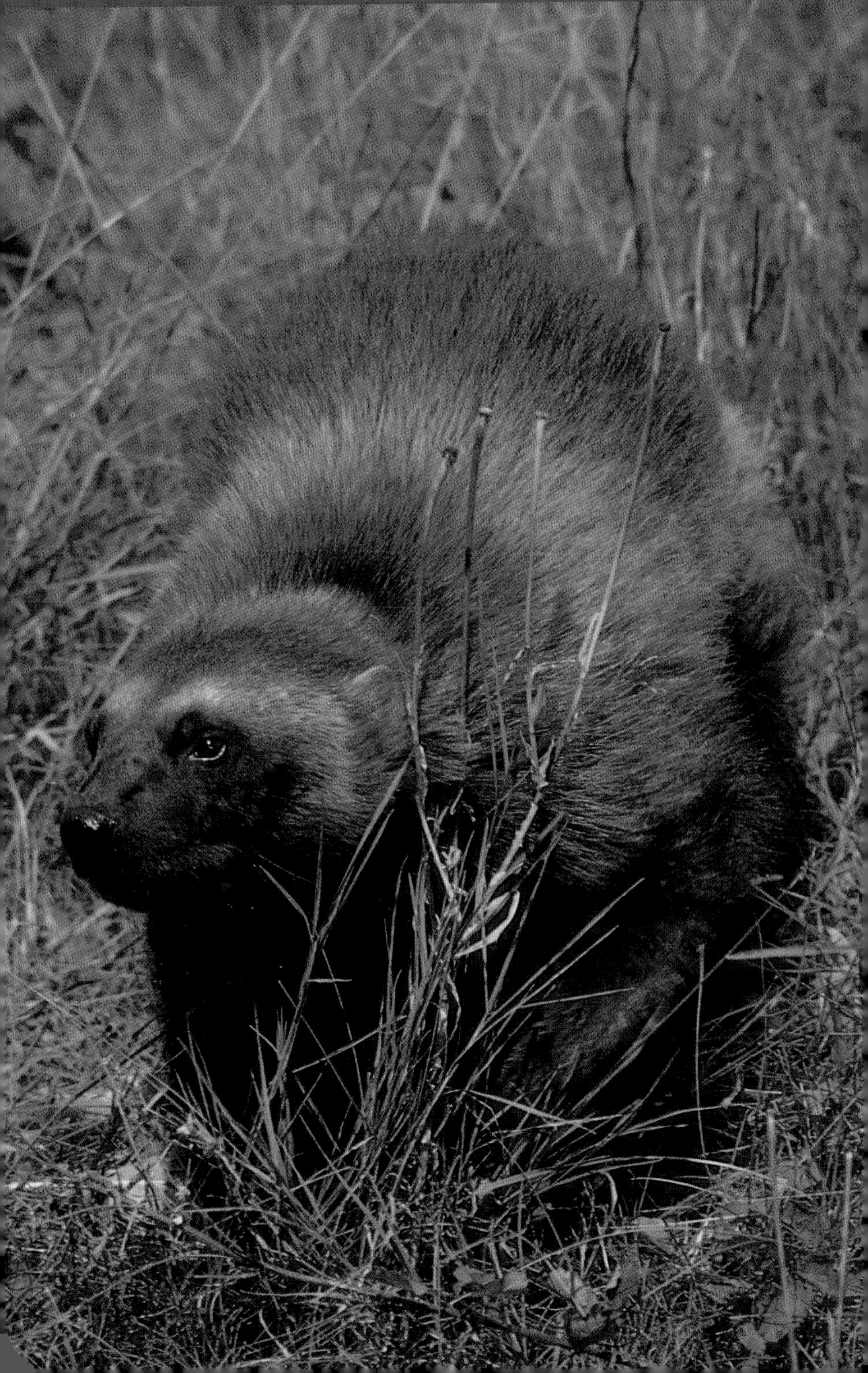

DEN: The den may be among the roots of a fallen tree, in a hollow tree butt, in a rocky crevice or even in a long-lasting snowbank. The natal den is often underground, and it is lined with leaves by the mother. A Wolverine may maintain several dens throughout its territory, ranging in quality from makeshift cover under tree branches to a permanent underground dugout.

YOUNG: Wolverines breed between late April and early September, but the embryos do not implant in the uterine wall until January. Between late February and mid-April, the female gives birth to a litter of one to five (generally two or three) cubs. The stubby-tailed cubs are born with their eyes and ears closed and with a fuzzy, white coat that sets off the darker paws and face mask. They nurse for eight to nine weeks; then they leave the den and their mother teaches them to hunt. The mother and her young typically stay together through the first winter. The young disperse when they become sexually mature in spring.

loping trail

SIMILAR SPECIES: The **Badger** (p. 80) is squatter, with a distinctive vertical stripe on its forehead and without the lighter side stripes of a Wolverine. The **Black Bear** (p. 102) is larger, has a shorter tail, and lacks the light buffy stripe along each side.

Badger

Badger
Taxidea taxus

In tall-grass prairies and open pastures, a lucky observer may encounter a Badger. With a flair for remodelling, Badgers are nature's rototillers and backhoes. The large holes left by Badgers are of critical importance as den sites, shelters and hibernacula for dozens of species, from Coyotes (p. 110) to Black-widow Spiders. Even toads can be found inside the cool, moist burrows.

The Badger enjoys a reputation for a fierceness and boldness that was acquired in part from a distantly related weasel bearing the same name in Europe. While it is true that a cornered Badger will put up an impressive show of attitude, like most animals it prefers to avoid a fight. When it is severely threatened or in competition, the Badger's claws, strong limbs and powerful jaws make this animal a dangerous opponent. In spite of its impressive arsenal, the Badger routinely kills only rodents and other small mammals. However, rare occasions are known where Badgers have taken Coyote pups. Conversely, a group of Coyotes may defeat a Badger.

Pigeon-toed and short-legged, the Badger is not much of a sprinter, but its heavy front claws enable it to move large quantities of earth in short order. Although a Badger's predatory nature is of benefit to landowners, its natural digging skills have led many Badgers to be killed—cattle and horses have been known (very rarely) to break their legs when stepping carelessly into Badger excavations.

Badgers tend to spend a great part of winter sleeping in their burrows, but they do not enter a full state of hibernation like their European relatives or like many of the small mammals upon which they feed. Instead, Badgers emerge from their slumber to hunt whenever winter temperatures are more moderate.

In spite of low population densities, almost all sexually mature female Badgers are impregnated during the nearly three months that they are sexually receptive. As with most members of the weasel family, once the egg is fertilized further embryonic development is put off until the embryos implant, usually in January, which will result in a spring birth.

DESCRIPTION: Long, grizzled, yellowish-grey hair covers this short-legged, muscular member of the weasel family. The hair is longer on the sides than on the back or belly, which adds to the flattened appearance of the body. A white stripe originates on the nose and runs

RANGE: From north-central Alberta and Saskatchewan, the Badger ranges to the southeast throughout the Great Plains and Prairies and southwest to Baja California and the central Mexican highlands.

Total Length: 80–85 cm
Tail Length: 13–16 cm
Weight: 5–11 kg

back onto the shoulders or sometimes slightly beyond. The top of the head is otherwise dark. A dark, vertical crescent or triangle is apparent on each side of the face between the short, rounded, furred ears and the eyes. A whitish or pale buff stripe originates at the edge of the mouth on each side and passes beside the eye and on up to just above the ear. The short, bottlebrush tail is more yellowish than the body, and the lower legs and feet are very dark brown, becoming blackish at the extremities. The three central claws on each forefoot are greatly elongated for digging. Skulls from older Badgers often have the "wrap-around" jaw articulation seen in Wolverine skulls.

HABITAT: Essentially an animal of open places, the Badger shuns thick forests. It is usually found in association with ground-dwelling rodents, typically in grasslands, agricultural areas and open parkland. In Ontario it is only found in open areas of the southwest and in western grasslands close to the Manitoba and Minnesota borders.

FOOD: Burrowing mammals fill most of the Badger's dietary needs, but it also eats eggs, young ground-nesting birds, mice and sometimes carrion, insects and snails.

DEN: A Badger may dig its own den or take over another animal's burrow. The den may approach 9 m in length and have a diameter of about 30 cm. The

DID YOU KNOW?

Badgers make an incredible variety of sounds: adults hiss, bark, scream and snarl; in play, young Badgers grunt, squeal, bark, meow, chirr and snuffle; and the front claws clatter when a Badger runs on a hard surface.

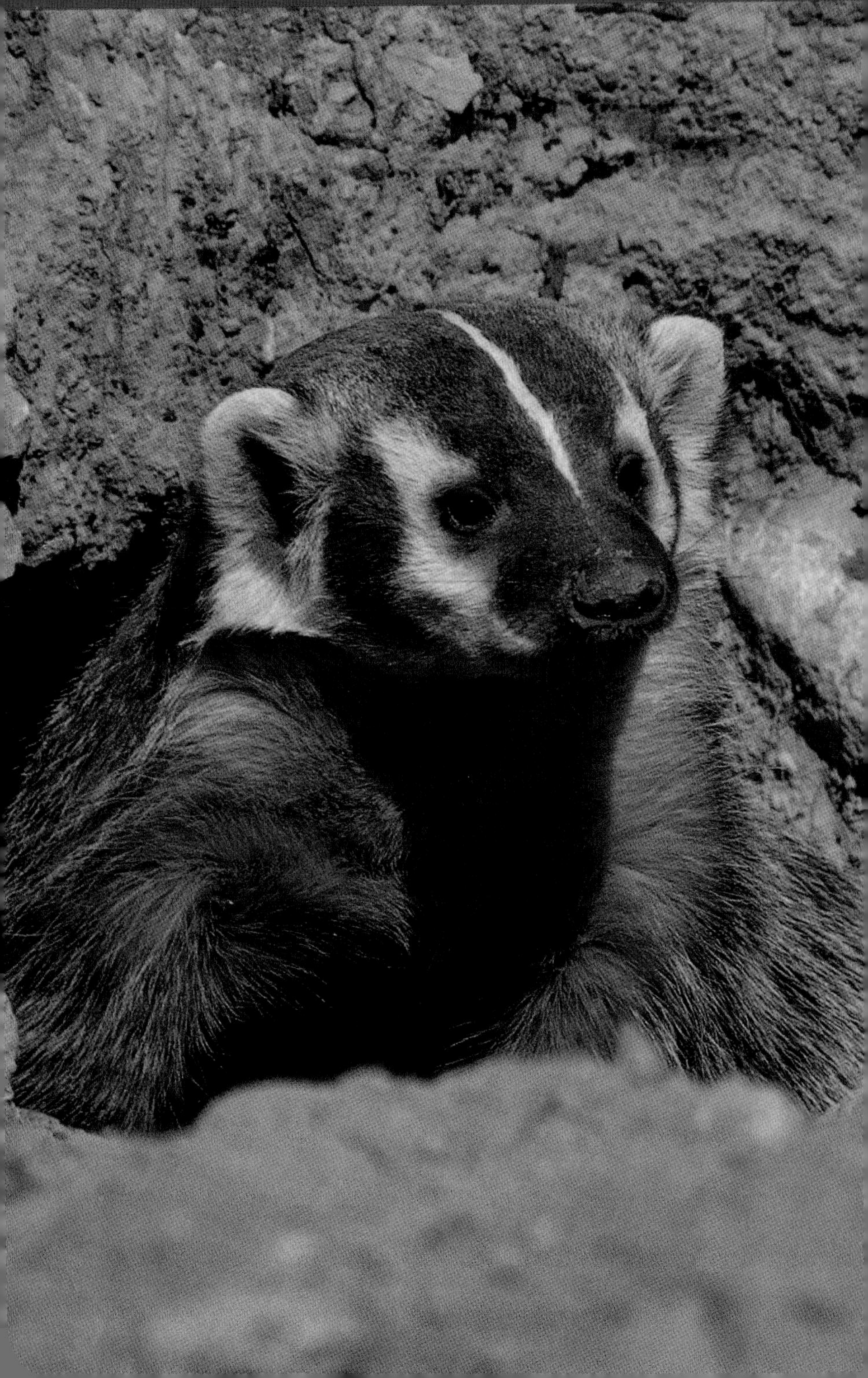

left foreprint

walking trail

Badger builds a bulky grass nest in an expanded chamber near or at the end of the burrow. A large pile of excavated earth is generally found to one side of the burrow entrance.

YOUNG: One to five (usually four) naked, helpless young are born between late April and mid-June. Their eyes open after a month, and at two months their mother teaches them to hunt. In early evening they leave the burrow, trailing their mother. The young investigate every grasshopper or beetle they encounter, but the mother directs the expedition to small mammal burrows. She often cripples a rodent and then leaves it for her young to kill. The young disperse in autumn, when they are three-quarters grown. Some of the young females may mate in their first summer, but most Badgers are not sexually mature until they are a year old. Delayed implantation of the embryo is characteristic.

Raccoon

SIMILAR SPECIES: The **Raccoon** (p. 88) has a long, ringed tail and horizontal black-and-white facial markings, rather than vertical markings like the Badger. The **Wolverine** (p. 76) might be confused with a Badger, but the two have different ranges and the Badger's facial pattern and overall appearance are unique. The **Striped Skunk** (p. 60) also has black-and-white markings, but it is smaller and has a mainly black body.

Northern River Otter

Lontra canadensis

It may seem too good to be true, but all those playful characterizations of the Northern River Otter are founded on truth. Otters often amuse themselves by rolling, sliding, diving or "body surfing," and they may also push and balance floating sticks with their noses or drop and retrieve pebbles for minutes at a time. They seem particularly interested in playing on slippery surfaces—they leap onto snow or mud with their forelegs folded close to their bodies for a streamlined toboggan ride. Unlike most members of the weasel family, river otters are social animals, and they will frolic together in the water and take turns sliding down banks.

With their streamlined bodies, rudder-like tails, webbed toes and valved ears and nostrils, river otters are well adapted for aquatic habitats. When they emerge from water to clamber over rocks, there is a serpentine appearance to their progression. The large amounts of playtime they seem to have results from their efficiency at catching prey when it is plentiful. Although otters generally cruise along slowly in the water by paddling with all four feet, they can dart after prey with the ease of a seal whenever hunger strikes. When an otter swims quickly, it propels itself mainly with vertical undulations of its body, hindlegs and tail. Otters can hold their breath for as long as five minutes.

Because of all its activity, the Northern River Otter leaves many signs of its presence when it occupies an area. Its slides are the most obvious and best-known evidence—but be careful not to mistake the slippery Beaver trails that are common around Beaver ponds for otter slides. Despite its other aquatic tendencies, an otter always defecates on land. Its scat is simple to identify—it is almost always full of fish bones and scales.

A river otter may make extensive journeys across land, even through deep snow. Although this animal looks clumsy on land, it can easily outrun a human with its humped, loping gait. On slippery surfaces, such as wet grass, snow and ice, the otter glides along, usually on its belly with its legs tucked either back or forward to help steer and push. On flat ground, snowslides are sometimes pitted with blurred footprints where the otter has given itself a push for momentum.

In the past, the Northern River Otter's thick, beautiful, durable fur led to excessive trapping that greatly diminished its continental population. Trapping has since been reduced, and the otter seems to be slowly recolonizing parts of North

RANGE: The Northern River Otter occurs from near treeline across Alaska and Canada south through forested regions to northern California and northern Utah in the West and Florida and the Gulf Coast in the East. It is largely absent from the Midwest and Great Plains.

Total Length: 0.9–1.4 m
Tail Length: 30–50 cm
Weight: 5–11 kg

America from which it has been absent for decades. Even in areas where it is known to occur, however, it is infrequently seen. Water quality can influence otter densities because much of the prey is aquatic.

DESCRIPTION: This large, weasel-like carnivore has dark brown upperparts that look black when wet. It is paler below, and the throat is often silver grey. The head is broad and flattened, and it has small eyes and ears and prominent, whitish whiskers. The feet are webbed. The long tail is thick at the base and gradually tapers to the tip. The male is the larger gender. The otter does not hibernate, and in winter it still chases fish under the surface of the ice.

HABITAT: Year-round, river otters live primarily in or along wooded rivers, ponds and lakes, but they sometimes roam far from water. They may be active day or night but tend to be more nocturnal close to human activity. In winter, Northern River Otters are found on lakes with beaver lodges or on bog ponds with steep banks containing old beaver burrows, through which they can enter the water. Also, riffles and waterfalls with pools provide important access to water in the winter.

DID YOU KNOW?

When a troupe of agile river otters travel single file through the water, their undulating, lithe bodies combine to form a very serpent-like image—perhaps with enough similarity to give rise to rumours of lake-dwelling sea-monsters.

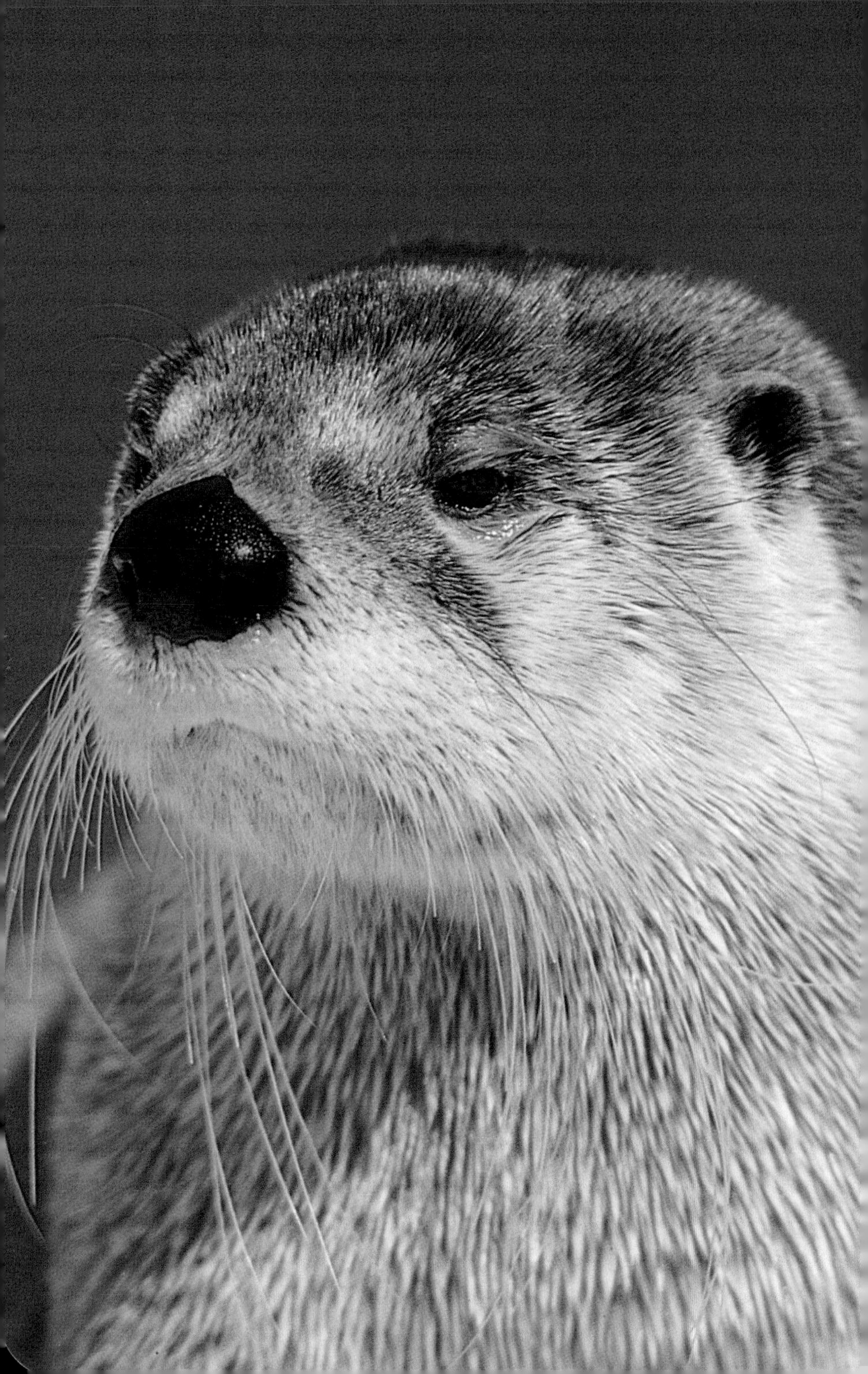

FOOD: Crayfish, turtles, frogs and fish form the bulk of the diet, but otters occasionally depredate bird nests and eat small mammals, such as mice, young Muskrats and young Beavers, and sometimes even insects and earthworms.

foreprint

DEN: The permanent den is often in a bank, with both underwater and above-water entrances. When roaming on land, an otter rests under roots or over-hangs, in hollow logs, in the abandoned burrows of other mammals or in abandoned Beaver or Muskrat lodges. Natal dens are usually abandoned Muskrat, Beaver or Woodchuck dens.

YOUNG: The female bears a litter of one to six blind, fully furred young in March or April. The young are 140 g at birth. They first leave the den at three to four months and leave their parents at six to seven months. Otters become sexually mature at two years. The mother breeds again soon after her litter is born, but delayed implantation of the embryos puts off the birth until the following spring.

loping trail

SIMILAR SPECIES: The **Beaver** (p. 156) is stouter and has a wide, flat, hairless, scaly tail. The **Mink** (p. 74) is much smaller, its feet are not webbed, and its shorter tail is cylindrical, not tapered.

Beaver

Raccoon
Procyon lotor

The Raccoon is famous for its black bandit mask and ringed tail. The mask suits the Raccoon, because it is well known as a looter of people's gardens, cabins, campsites and, yes, even garbage cans. A Raccoon is likely to investigate tasty bits of food and any shiny object it finds. Despite its roguish behaviour, however, the Raccoon has never been associated with ferociousness or savagery—it is mainly a curious and docile animal unless it is cornered or threatened. Testing a Raccoon's ferocity is an unnecessary and simple-minded act, and Raccoons have been known to wound and even kill attacking dogs.

Raccoons are among the most frequently encountered wild carnivores in many parts of Ontario. When Raccoons are seen, which is usually at night, they quickly bound away, effectively evading flashlight beams by slipping into burrows or climbing to tree retreats. Should their tree sanctuary be found, Raccoons remain still at a safe distance, waiting for the invasive experience to end.

Raccoons tend to frequent muddy environments, a characteristic that allows people to find their diagnostic tracks along the edges of wetlands and waterbodies. Like bears (pp. 102–09) and humans, Raccoons walk on their heels, so they leave unusually large tracks for their body size. They will methodically circumnavigate wetlands in the hopes of finding duck nests or unwary amphibians upon which to dine. The way that the Raccoon typically feels its way through the world has long been recognized. In fact, our word "raccoon" is derived from the Algonquian name for this animal, *aroughcoune,* which means, "he scratches with his hands." One of the best-known characteristics of the Raccoon is its habit of dunking its food in water before eating it. It had long been thought that the Raccoon was washing its food—the scientific name *lotor* is Latin for "washer"—but biologists now believe that a Raccoon's sense of touch is enhanced by water, and that it is actually feeling for inedible bits to discard.

Long, cold winters are an ecological barrier to the dispersal of this animal, because it does not hibernate and so requires year-round food availability. It may sleep for extended periods in parts of its range, but it still comes out on warmer nights. Over the past century, however, the Raccoon has been moving into colder climes, perhaps because of increasing human habitation in these areas. When Raccoons first appeared in Winnipeg, Manitoba, in the 1950s, many people were quite surprised and

RANGE: The Raccoon occurs from southern Canada south through most of the U.S. and Mexico. It is absent from parts of the Rocky Mountains, central Nevada, Utah and Arizona.

Total Length: 65–100 cm
Tail Length: 19–40 cm
Weight: 5–14 kg

took them to the local zoo, thinking they were escapees rather than a new species of urban "wildlife."

DESCRIPTION: The coat is blackish to brownish grey overall, with lighter, greyish-brown underparts. The bushy tail, with its four to six alternating blackish rings on a yellowish-white background, makes the Raccoon one of the most recognizable North American carnivores. There is a black "mask" across the eyes, bordered by white "eyebrows" and mostly white snout, and a strip of white fur separates the upper lip from the nose. The ears are relatively small. Raccoons are capable of producing a wide variety of vocalizations: they can purr, growl, snarl, scream, hiss, trill, whinny and whimper.

HABITAT: Raccoons are most often found near streams, lakes and ponds. They favour woodlands and are not found in vast open grasslands or tundra.

FOOD: The Raccoon fills the role of medium-sized omnivore in the food web. Besides eating fruits, nuts, berries and insects, it avidly eats clams, frogs, fish, eggs, young birds and rodents. Just as a bear does, the Raccoon consumes large amounts of food in autumn to build a large fat reserve that will help sustain it over winter.

DEN: The den is often located in a hollow tree, but Raccoons are increasingly using sites beneath abandoned buildings or under discarded construction materials. In rougher terrain, dens can sometimes be found in rock crevices, where

DID YOU KNOW?

Raccoons have thousands of nerve endings in their "hands" and "fingers." It is an asset they constantly put to use, probing under rocks and in crevices for food.

right foreprint

walking trail

grasses or leaves carried in by the female may cover the floor.

YOUNG: After about a two-month gestation, the female bears two to seven (typically four) young in late spring. The young weigh just 57 g at birth. Their eyes open at about three weeks, and when they are six to seven weeks old they begin to feed outside the den. At first, the mother carries her young about by the nape of the neck, as a cat carries kittens. About a month later, she starts taking them on extended nightly feeding forays. Some young disperse in autumn, but others remain until their mother forces them out when she needs room for her next litter.

Badger

SIMILAR SPECIES: Only the **Badger** (p. 80) could possibly be confused with a Raccoon, but a Badger is much squatter; its facial markings are vertically oriented, unlike the horizontal "mask" of a Raccoon; and its shorter, thinner tail does not have the Raccoon's distinctive rings.

Walrus

Odobenus rosmarus

Truly one of a kind, the Walrus is the sole member of its family. It most likely evolved from an ancestral eared seal, but 7 million years of adaptation has made the Walrus unlike any other living seal. The Walrus's unique tusks, which grow throughout an individual's life, are its most distinguishing feature. The tusks are modified teeth, and they are coated in enamel when they first erupt. Over time, the enamel wears off, leaving a tusk of pure ivory.

The exploitation of Walruses for their ivory was once extensive, but it has nearly been stopped, thanks to the Canadian and U.S. governments' involvement. Only the Inuit, who have traditional links to this animal, continue to hunt the Walrus. The Inuit use all parts of the animal: the meat is eaten; the bones become tools; the skin becomes leather for clothing or covers for boats; the blubber is used for oil and fuel; the intestines are used to make waterproof containers; and the tusks are used as sled runners or for ornamentation. Recently, the Inuit have made ivory carvings to bring additional revenue into their communities.

There are two distinct Walrus races: the Pacific Walrus, which is the larger of the two, inhabits waters from the Chukchi Sea off northeastern Siberia to Alaska; the Atlantic Walrus lives in Hudson Bay and the Arctic seas to Greenland. Both races of Walrus live in groups of anywhere from a few dozen to up to 2000 individuals. The smaller Atlantic Walrus is not migratory, unlike its Pacific counterpart, and it hauls out either on land or landfast ice that does not travel.

The Walrus is an excellent swimmer and can reach speeds up to 24 km/h. When it dives for food, it can descend to 91 m, remaining underwater for as long as 30 minutes.

DESCRIPTION: Both bull and cow Walruses have tusks, but a cow's tusks are considerably shorter and more curled than a bull's. Walruses have very little hair over their bodies, and their skin colour varies from pale pink, yellow, reddish or brownish to nearly white. The hind flippers have nails on all five digits and can be rotated underneath so the walrus can "walk" on all fours on land. The fore flippers lack nails completely. The face is covered with about 400 whiskers, each up to 30 cm in length. Walruses have no visible tail.

HABITAT: Historically, Walruses may have inhabited more of the rocky coasts

RANGE: The Atlantic race of the Walrus is found in the Arctic seas from Greenland to northern Canada, south into Hudson Bay and rarely along Labrador.

Total Length: male up to 3.7 m; female up to 2.9 m
Weight: male up to 1000 kg; female up to 800 kg
Tusk Length: male up to 76 cm; female up to 36 cm

and islets of mainland Canada, including Hudson Bay, than they do now, but the impact of human hunting has pushed them farther into the Arctic seas. Currently, they spend most of their time on Arctic pack ice over water that is about 18 m deep. Walruses spend more of their time basking than do other seals, usually on ice or on rocky outcroppings.

FOOD: Using its mouth like a powerful vacuum, the Walrus sucks mollusks out of their shells. Clams are the most common food, and a large hungry Walrus can polish off up to 6000 of these bivalves in one day. The Walrus is opportunistic, and it will feed on fish, crabs, shrimp, worms and a variety of other marine creatures.

YOUNG: Unlike other seals that usually mate each year, a female Walrus mates once every two years. Gestation is 11 to 12 months, and the single newborn calf weighs 45–68 kg. The cow nurses her calf in the water in a vertical position. The calf remains with the mother and nurses from her for nearly two years. The mother separates from the calf before she has another baby.

SIMILAR SPECIES: No other seal grows tusks, and both the **Harbour Seal** (p. 94) and **Ringed Seal** (p. 98) are much smaller. A large **Bearded Seal** (p. 102) can be as long as a Walrus, but a Walrus is much more rotund.

DID YOU KNOW?

Although Walruses may sleep on land, they have the ability to sleep vertically in the water with their heads held up by inflated air sacs in the neck.

Harbour Seal
Phoca vitulina

The inquisitive Harbour Seal is a well-known resident of the coastal waters elsewhere in Canada, but it is extremely rare on the Ontario coast. The few confirmed records are from before 1900 along the Ottawa and St. Lawrence rivers.

Harbour Seals are frequently referred to as sociable or gregarious, but this perception is not entirely true. Although many seals may bask on rocks together, they pay very little attention to their neighbours and seldom interact. Only during the pupping season is there interaction, and it is primarily between females. Mothers with newborn pups may congregate in a "nursery" in shallow water where the pups can sleep. These nursery groups are not interactive; they form solely as a protective measure against possible predation. While most of the females and pups sleep, some are likely to be awake and watchful for danger. The same is true for hauled-out seals. Where several seals are together, chances are good that there is always at least one individual awake and wary of danger.

Harbour Seals cannot sleep at the surface like sea-lions and sea otters. During the day, they can sleep underwater in shallow coastal water by resting vertically just above the bottom. Young pups commonly rest in this manner. They can go without breathing for nearly 30 minutes, and though they sometimes wake up to breathe, they frequently rise to the surface and take a breath without awakening, and then sink back to the bottom. At night when the tide is out, they sleep high and dry in their preferred haul-out site. They frequently rest with their heads and rear flippers lifted above the rock.

The Harbour Seal tends to be wary of humans, and if you approach one it is likely to dive immediately into the water. On the other hand, many kayakers and boaters have enjoyed watching inquisitive individuals that approach their boats for a better look. This kind of experience is controlled by the seal; if it wants to see you, it will come closer. If the seal is afraid of you, it will leave. Do not approach a seal that has tried to flee you because it can cause unnecessary stress to the animal.

DESCRIPTION: A Harbour Seal is typically dark grey or brownish grey with light, blotchy spots or rings. The reverse colour pattern is also common—light grey or nearly white with dark spots. The undersides are generally lighter than the back. The outer coat is composed of stiff guard hairs about 1 cm

RANGE: Harbour Seals are found along the western coast of North America from northern Alaska to California and in the East from southern Baffin Island and Hudson Bay to the Carolinas. They also inhabit parts of Europe and Asia.

Total Length: 1.2–1.8 m
Tail Length: 9–11 cm
Total Weight: 50–140 kg

long, and this characteristic is what gives seals in this family the name "hair seals." The guard hairs cover an insulating undercoat of sparse, curly hair that is about 0.5 cm long. Pups bear a spotted, silvery or grey-brown coat at birth. The head is large and round, and there are no visible ears. Each of the short front flippers bears long, narrow claws. The male is the larger gender in this species.

HABITAT: In other parts of Canada, this nearshore species is frequently found in bays and estuaries. Common haul-out sites include intertidal sandbars, rocks and rocky shores, and favoured spots

are used by Harbour Seals generation after generation.

FOOD: Harbour Seals feed primarily on fish, such as rockfish, cod, herring, flounder and salmon. To a lesser extent, they also feed on mollusks, such as clams, squid and octopus, and crustaceans, such as crayfish, shrimp and

DID YOU KNOW?

Sometimes these seals follow fish several hundred kilometres up major rivers; there are even residents in some inland lakes.

crabs. Newly weaned pups seem to consume more shrimp and mollusks than do adults. Adult Harbour Seals have been seen taking fish from nets and some have even entered fish traps to feed, making a clean getaway afterwards.

YOUNG: The breeding season for Harbour Seals varies geographically. The farther north a population, the later the breeding and pupping. Gestation lasts 10 months, and a single pup is born between April and August. The pups are weaned when four to six weeks old—after they have tripled their birth weight on their mother's milk, which is more than 50 percent fat. Within a few days of weaning their pups, females mate again. Harbour Seals become sexually mature at three to seven years. Captive seals have lived more than 35 years, although the typical life span in the wild is 25 years.

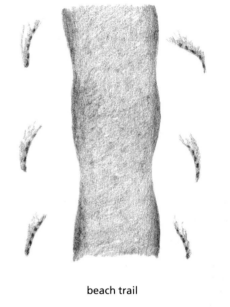

beach trail

Ringed Seal

SIMILAR SPECIES: The **Ringed Seal** (p. 98) is smaller and has irregular light rings on its sides. The **Bearded Seal** (p. 102) is larger and uniformly coloured and bears a distinct "beard" of long whiskers. The **Walrus** (p. 92) is much larger and has tusks.

Ringed Seal
Pusa hispida

Studies of the Ringed Seal have provided information not only about the biology of this small seal, but also about the biogeological history of the continent. Fossils of Ringed Seals have been found near Ottawa, and we know that seas covered this region more than 10,000 years ago. By identifying the variety of sea creatures Ringed Seals feed on today, we can draw inferences about the animals living in those ancient seas that once covered the province.

When a Ringed Seal dives for crustaceans, it can reach a maximum depth of about 91 m. It can stay under for up to 18 minutes, though a dive usually lasts for five minutes or less. Unlike its cousin the Bearded Seal (p. 100), the Ringed Seal cannot tolerate the pressure at great sea depths, nor can it tolerate being without air for much longer than 15 minutes. As a result, the Ringed Seal remains in shallower areas, while the Bearded Seal occupies deeper waters.

Although it spends most of its time in the water, the Ringed Seal digs small chambers under the snow, but above the ice, where it rests when it needs to reoxygenate its blood after a long dive. Sometimes, many seals are found together in chambers that are connected by short passageways.

The Walrus (p. 92) is the biggest threat to the Ringed Seal in the water, while the Polar Bear (p. 106), the Arctic Fox (p. 118) and humans are the dangerous creatures on land and ice. There is no place where this seal is truly safe, and it seems to know it—every minute or so a Ringed Seal lifts its head to scan its surroundings for danger. Even in its snowy den, this seal is watchful and ready to flee a Polar Bear that has sniffed it out and dug through the snow to reach it. The Arctic Fox, which preys on the pups, is small enough to walk right into the den.

DESCRIPTION: One of the smallest aquatic carnivores, this little seal reaches a maximum length of 1.7 m, though it is usually only 1.2 m long. Its fur is variegated brown or even bluish black above and pale yellowish or grey below, with irregular doughnut-shaped rings of light-coloured fur on the sides and back. It has long whiskers and eyebrows, a short muzzle and dark eyes. Males are typically larger than females.

HABITAT: The Ringed Seal spends most of its life underwater in the northern polar seas. It maintains breathing holes under the ice by digging the ice away

RANGE: Ringed Seals are found in the Arctic seas from Point Barrow, Alaska, in the West to Hudson Bay, Labrador and Newfoundland in the East.

Total Length: 1.2–1.7 m
Tail Length: 8 cm
Weight: 50–110 kg

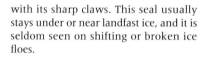

with its sharp claws. This seal usually stays under or near landfast ice, and it is seldom seen on shifting or broken ice floes.

FOOD: In shallow water these seals feed mainly on shrimp, small fish and crustaceans. In deep water their diet shifts to larger zooplankton, such as amphipods.

YOUNG: A female typically mates annually, in April or early May, and gives birth to a single pup the following March, after 11 months of gestation. The Ringed Seal is unique among seals because of the maternal den it constructs under the snow. A pregnant female may hollow out a shallow chamber in the snow above a breathing hole, or she may use a natural snow cave. Maternal chambers are typically about 2 x 3 m wide and only 60 cm deep. The young are born helpless, but with all their teeth and covered in lanugo (the soft, downy fur that is reminiscent of a white teddy bear). A pup nurses for about two months. When its mother leaves the den to feed, the pup remains by itself and waits quietly. Some pups may stay near their mothers for the remainder of the year, while other pups may disperse soon after weaning.

SIMILAR SPECIES: The **Harbour Seal** (p. 94) is slightly larger, and its marking are spotty or patchy rather than the distinct "doughnut" markings on the Ringed Seal. The **Bearded Seal** (p. 100) is larger and uniformly coloured and bears a distinct "beard" of long whiskers. The **Walrus** (p. 92) is much larger and has tusks.

DID YOU KNOW?

The presence of PCBs and DDT in the environment is believed to have reduced the pregnancy rates of female Ringed Seals to as low as 28 percent in certain areas—a percentage low enough to threaten the survival of some local Ringed Seal populations.

Bearded Seal
Erignathus barbatus

The Bearded Seal is perhaps the most vocal of all seals—what it lacks in its nondescript appearance, it easily makes up for with its repertoire of unusual and haunting calls. In spring, the male Bearded Seal sings melodious yet eerie songs beneath the ice. Each song can last for up to a minute, and they carry for a long distance underwater. To court a female, a bull hauls out on the ice and serenades the cows with his echoing, warbling calls.

When humans hold their breath, the increasing carbon dioxide levels in the blood trigger them to breathe again in a very short time. This trigger is greatly desensitized in seals, and they are able to postpone breathing for about 30 minutes. When a seal is underwater, its nostrils are closed and its heartbeat is slowed from the normal 100 beats a minute to about 10 beats a minute. During the dive, the muscles function mainly anaerobically, creating an oxygen debt in the body. Seals store oxygen in their blood and in certain molecules of the muscle itself, and this stored oxygen is used, or "spent," during the dive. The longer the dive lasts, the greater the body's oxygen debt becomes. When the dive is over, the seal must rest and rebuild, or "pay back," the oxygen

stores of its body. In other words, when a Bearded Seal is basking, it is not just enjoying the sun, it is allowing its body to slowly reoxygenate after a long dive.

DESCRIPTION: As its name suggests, this seal has a beard of long, flat bristles on either side of the muzzle. Other seals have whiskers, but none compare to the long, bristly beard of the Bearded Seal. The head of this large seal is noticeably small for its body size. The fur varies in colour from uniformly light brown to tawny golden to greyish. The front flippers are distinctively squarish, and the third digit is longer than all the rest. In a switch from many other pinnipeds, female Bearded Seals are slightly larger than males.

HABITAT: The Bearded Seal spends most of its time in the water of cold Arctic seas. It primarily stays in water no deeper than 150 m, because it dives to the bottom to feed. The Bearded Seal is only rarely seen on land—it prefers open water and rests on drift ice and ice floes. This seal is not migratory, but it does follow the ice as it moves seasonally.

FOOD: Using its powerful claws and keenly sensitive whiskers, the Bearded Seal locates bottom-dwelling fish and

RANGE: Bearded Seals inhabit the northern coastal waters and shallow Arctic seas from Point Barrow, Alaska, to Labrador.

Total Length: 2.1–3 m
Tail Length: 10–15 cm
Weight: 200–350 kg

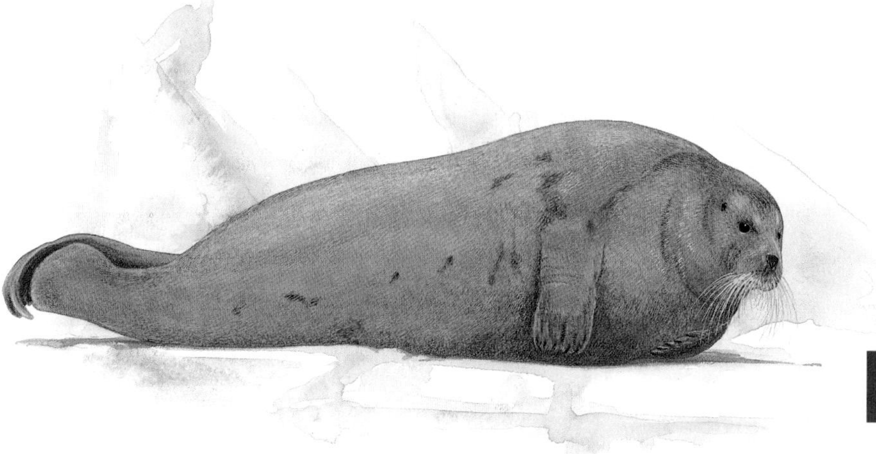

shellfish to eat. Favourite foods include cod, sculpin, flounder, octopus and a variety of shellfish, such as whelks and clams.

YOUNG: In May, the mating season is marked by these normally solitary seals forming groups of up to 50 individuals. The males are marvellously vocal, and they emit warbling calls on the ice surface to serenade the cows. Females are not vocal. Sometimes the bulls fight each other to establish hierarchy, but the dominant bulls do not form harems. A female gives birth on the ice to one young in April or May, after a gestation of about 11 months. The pup is well developed at birth and is nursed for no more than 18 days. The pup puts on blubber during this short nursing period and goes from weighing 32 kg at birth to about 91 kg by day 18, at which time the pup is weaned and the mother leaves.

SIMILAR SPECIES: The **Harbour Seal** (p. 94) is much smaller, and it has spotty or patchy markings. The **Ringed Seal** (p. 98) is also much smaller, and it has distinct "doughnut" markings on its sides. The **Walrus** (p. 92) is much heavier and has tusks.

DID YOU KNOW?

Each long bristle of the Bearded Seal's "beard" is extremely sensitive, and they help the seal locate its favourite foods in muddy sea bottoms.

Black Bear

Ursus americanus

The Black Bear, an inhabitant of forests and open, marshy woodlands throughout Ontario, is often feared by city dwellers who come to wilderness parks to appreciate the scenery. People who are more experienced with the wild forest and with animal behaviour may not fear the Black Bear but still treat it with healthy respect.

Contrary to popular belief and their classification as carnivores, Black Bears do not readily hunt larger animals. They are primarily opportunistic foragers and feed on what is easy and abundant—usually berries, horsetails, other vegetation and insects, although they will not turn up their noses at fish, young fawns or another carnivore's kill. Black Bear sows with young cubs are likely to attack young or sickly Moose (p. 34), deer and Elk (p. 26), but on the whole, large males are the more predatory of the two sexes.

In the past few decades, the ubiquitous dandelion has become increasingly abundant along roadsides and swaths cut into the forests, especially in central interior regions. In pursuit of its new favourite food, the Black Bear is now more frequently seen along roadsides. With dandelion leaves sticking out of its mouth and the puffy seeds stuck over its face and muzzle, the bear looks like a little kid covered in its favourite ice cream. Unfortunately, together with an increase in bear sightings along roadsides, vehicle collisions that may claim bears' lives are also increasing.

Within its territory, a bear has favourite feeding places and follows well-travelled paths to these sites. Keep in mind that the trails you hike in parks and wilderness areas may be used not only by humans, but also by bears en route to lush meadows or rich berry patches.

Normally, Black Bears are reclusive animals that will flee to avoid contact with humans if they hear you coming. If you surprise a bear, however, back away slowly. In particular, heed its warning of a foot stamp, a throaty "huff" or the champing sound of its teeth. The bear is agitated and probably does not like you, and it is giving you a clear warning to retreat from its territory. Research suggests that a bear may charge or attack when these warning signals are not understood by a person who instead remains frozen in place. The bear likely interprets such behaviour as a challenge.

Mismanagement of food and garbage in areas where bears and people occur is the unfortunate cause of many bear deaths. Those bears that become habituated to people and their garbage often become nuisances and are destroyed.

RANGE: Across North America, the Black Bear occurs nearly everywhere there are forests, swamps or shrub thickets. It avoids grasslands and areas of dense human habitation.

Total Length: 1.4–1.8 m
Shoulder Height: 90–110 cm
Tail Length: 8–18 cm
Weight: 40–270 kg

A grim threat to Black Bears throughout the world is the illegal trade in body parts. Bear paws and gall bladders have high black-market values, and poaching occurs in both Canada and the U.S., but fortunately to a lesser extent than elsewhere. As populations of many bears around the world shrink, however, North American bears face an increasing threat. With bear numbers dwindling and black-market values increasing, it is feared that the generally well-protected animals in Canadian parks may become prime targets of the trade.

DESCRIPTION: The coat is long and shaggy and ranges in colour from black to honey brown. The body is relatively short and stout, and the legs are short and powerful. The large, wide feet have curved, black claws. The head is large and has a straight profile. The eyes are small, and the ears are short, rounded and erect. The tail is very short. An adult male is about 20 percent larger than a female.

HABITAT: Black Bears are primarily forest animals, and their sharp, curved fore-claws enable them to easily climb trees, even as adults. In spring, they often forage in natural or roadside clearings.

FOOD: In the wild, up to 95 percent of the Black Bear's diet is plant material: leaves, buds, flowers, berries, fruits and roots are all consumed. This omnivore also eats animal matter, such as small mammals, small vertebrates and many kinds of invertebrates, including bees (and their honey) and other insects; even young hoofed mammals may be killed and eaten. Carrion and human garbage are eagerly sought out.

DEN: The den, which is only used during winter, may be in a cave, beneath a fallen log, under the roots of a wind-thrown tree, or even in a haystack.

DID YOU KNOW?

During its winter slumber, a Black Bear loses 20 to 40 percent of its body weight. To prepare for winter, the bear must eat thousands of calories a day during late summer and autumn.

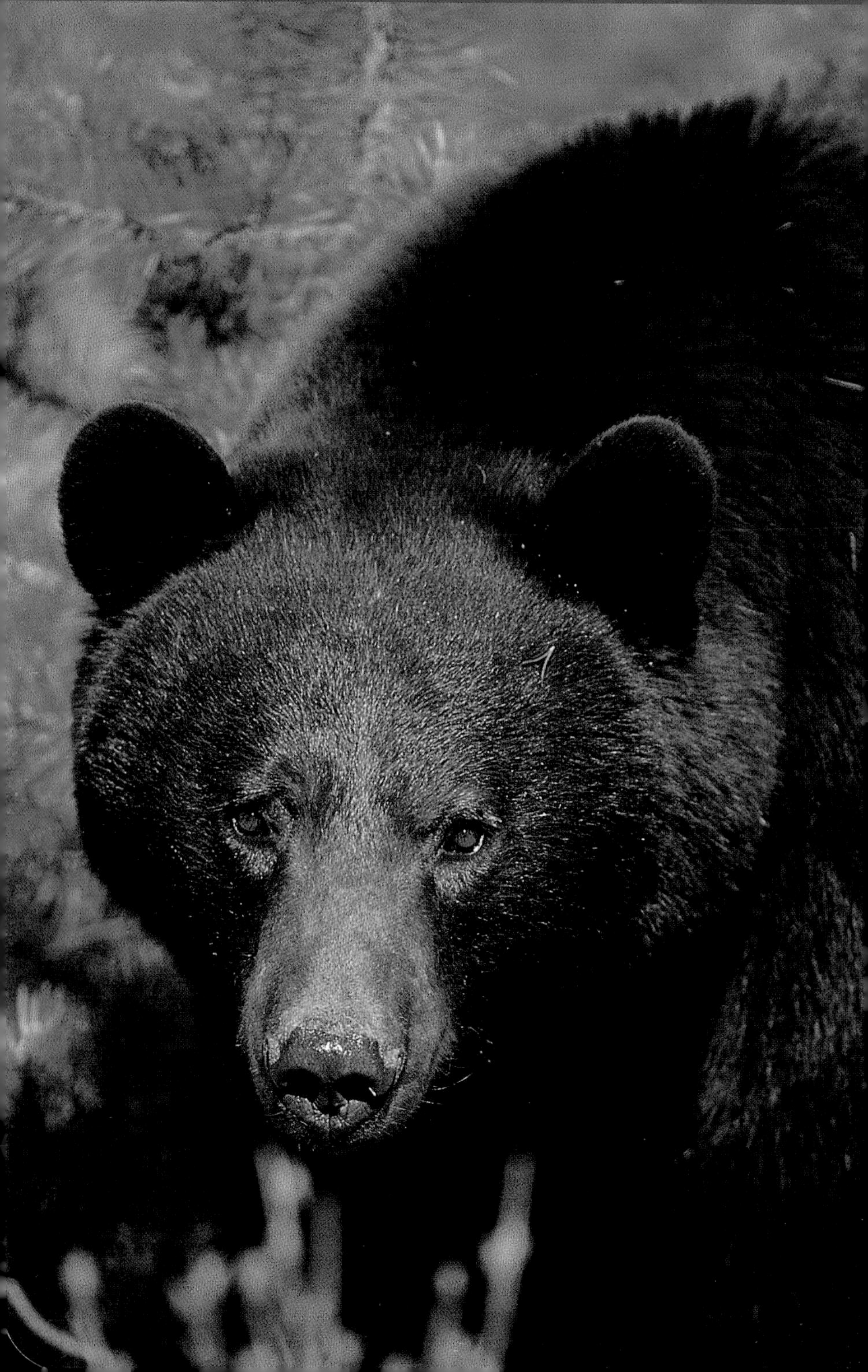

foreprint

walking trail

Some bears may even choose a depression in the ground and cover themselves with leaves. The bear will not eat, drink, urinate or defecate during its time in the den. The hibernation is not deep; instead, it is as if the bear is very groggy or heavily drugged. This state is called "torpor" and is technically not a true hibernation. Rarely, a bear may rouse from this torpor and leave its den on mild winter days.

YOUNG: Black Bears mate in June or July, but the embryos do not implant and begin to develop until the sow enters her den in November. The number of eggs that implant seems to be correlated with the female's weight and condition—fat mothers have more cubs. One to five (usually two or three) young are born in January, and they nurse while the sow sleeps. The cubs' eyes open, and they become active when they are five to six weeks old. They leave the den with their mother when they weigh 2–3 kg, usually in April. The sow and her cubs generally spend the next winter together in the den, dispersing the following spring. Black Bears typically bear young in alternate years.

SIMILAR SPECIES: The **Wolverine** (p. 76) looks a little like a small Black Bear, but it has a long tail, an arched back and pale side stripes. The larger **Polar Bear** (p. 106) is unmistakable because of its whitish coat.

Wolverine

Polar Bear

Ursus maritimus

The enormous Polar Bear, a symbol of the Canadian Arctic, is a uniquely adapted bear that is ideally suited for life in the frozen northern regions. The whitish, waterproof fur is hollow, and it acts like optical cables by transferring the energy from the sun to the skin. The Polar Bear's black skin maximizes absorption of this heat energy, and when seen through a heat-sensing device, the bear appears as a "cold" spot rather than the typical mammalian "hot" spot. Polar Bears are so well-adapted to cold that when individuals are on land in Ontario during the summer, they often use their winter dens again for a reprieve from the warmth of the summer (many winter dens are dug into the permafrost layer, and therefore stay cool in summer). This behaviour is only observable along the shores of Hudson Bay and James Bay, the largest known region for denning Polar Bears, and the only area where Polar Bears dig dens in land rather than ice.

Strictly speaking, the Polar Bear does not hibernate, even though it makes a winter den. The duration spent in the den is quite short and variable. Females den up for longer than the males, primarily because they give birth in January. Males may den up intermittently from November to January, but Polar Bears are known to be active at any time of the year. Unlike the other two bears in North America—the Black Bear (p. 102) and the Grizzly Bear (*U. arctos*)—the Polar Bear's diet is almost strictly carnivorous. When in the water, the Polar Bear has a swimming style unlike any other North American four-legged animal. It is able to reach speeds of nearly 10 km/h, and it does so by using only its forelegs. The hind legs, apparently unnecessary for propulsion, trail behind and are probably used like a rudder.

Today, Polar Bears are an internationally protected species. Much time and research went in to identifying their status and determining which areas are critical to their reproduction and survival. Since the livelihood of many northern indigenous peoples depends on these bears, many people are still permitted to hunt them. The number of bears hunted each year in each region is carefully controlled. The Inuit, the Greenlanders and some people in northern Russia take small numbers of Polar Bears each season.

DESCRIPTION: It is a common misconception that Polar Bears are white. Typically, only the new fur after a moult is white. Most Polar Bears are yellowish or

RANGE: Polar Bears have a northern circumpolar distribution. They are rarely found north of 88° N, and their southern limit is determined by pack and landfast ice in winter. Sometimes they are found as far south as the Pribilof Islands in the West and Newfoundland in the East.

Total Length: 200–335 cm
Tail Length: 8–13 cm
Height at Shoulder: 122–160 cm
Weight: males 300–800 kg; females 150–300 kg

greyish. Rare individuals are almost beige. Their coloration depends mainly on the season and light conditions. Seasonal moults change the fur colour, and exposure to sunlight oxidizes the coat and turns it yellow or tawny. These bears have black noses, lips and eyes, and under the fur their skin is also black. Relative to other bears, Polar Bears have long necks and long legs. Their ears are very small and rounded, and their tails are short and triangular.

HABITAT: In their Arctic home, Polar Bears live on the massive broken ice packs that characterize the region. Very seldom do these animals venture inland. Pregnant females visit coastline habitats when they are searching for suitable places to make a den. The Ontario portion of Hudson Bay provides excellent opportunities for viewing these bears.

FOOD: Active at any time of the day, Polar Bears feed primarily on seals and their pups. During the spring, more than half of their diet is Ringed Seal pups (p. 98). Other species of seals are eaten as well. Polar Bears will scavenge whale and Walrus carcasses when available. It is unlikely that Polar Bears will

DID YOU KNOW?

Many people believe that a Polar Bear covers its black nose with a paw while stalking prey to better camouflage itself. Although a plausible and intriguing theory, it has never been shown that Polar Bears actually act this way.

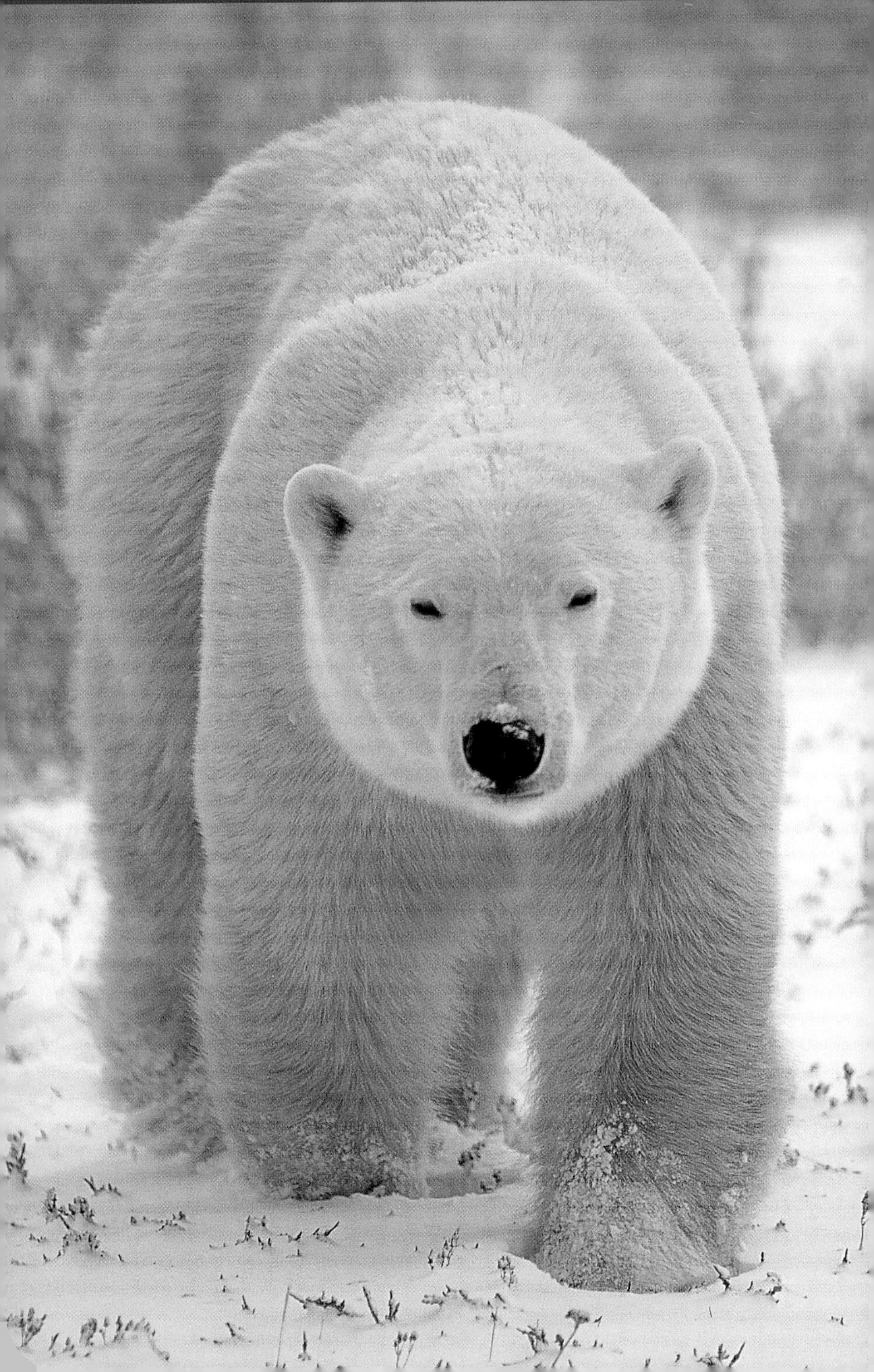

left foreprint

walking trail

hunt a Walrus (p. 92), because the two are nearly equally matched in battle.

DEN: In Ontario, pregnant females and some males make dens on shore, usually within 8 km of the water. The den may be dug into a slope where snow has accumulated up to 3 m deep, but it is frequently dug into a bank or peat hummock. Den design ranges from very simple to quite complex. A simple den is a single chamber with a short tunnel, and a complex den has several chambers and connecting tunnels. During the rest of the year on the ice packs, the bears may make a shelter in the snow to wait out inclement cold weather.

YOUNG: Mating occurs in April or May, but any one female only mates every two or three years. Two young is common, but only in years of favourable conditions will both young survive. The young are usually born in January. Cubs grow slowly and are not weaned until they are about two years old. When the cubs are one or two years, the female mates again, so that by the time the young are weaned and have dispersed she will give birth again.

SIMILAR SPECIES: Black Bears (p. 102) may produce nearly white individuals, but they are much smaller than Polar Bears. Although their ranges may overlap in northern Ontario, it is unlikely to find them in the same habitat.

Black Bear

Coyote

Canis latrans

A chorus of yaps, whines, barks and howls complements the evening and morning twilight hours throughout much of southern and central Ontario. Although Coyote calls are most intense during late winter and spring, corresponding to courtship, these excited sounds can be heard during suitable weather at any time of the day or year. Often initiated by one animal, several family groups may join in the ruckus until a large area is alive with their energetic calls.

Two centuries ago, the early explorers of this continent made frequent references in their journals to foxes and wolves, but they seldom mentioned Coyotes. Coyotes have increased their numbers across North America in the past century in response to the expansion of agriculture and forestry and the reduction of wolf populations. Despite widespread human efforts to exterminate them, they have thrived.

One of the few natural checks on Coyote abundance seems to be the Grey Wolf (p. 114). As the much larger and more powerful canids of the wilderness neighbourhood, wolves typically exclude Coyotes from their territories. Prior to the 19th century, the natural condition favoured wolves, but changes in the region since then have greatly

benefited Coyotes. Wolves prefer thick, unfragmented forests, while Coyotes are most common in open forests and grasslands. With increased fragmentation of Ontario's forests, Coyote populations have been increasing. Coyotes are now so widely distributed and comfortable with human development that almost all rural areas hold healthy populations, and even cities may have resident "urban coyotes."

Because of its relatively small size, the Coyote typically preys on small animals, such as mice, voles, ground squirrels, birds and hares, but it has also been known to kill deer, particularly their young. In Ontario, Coyotes generally form family-group packs, and they tend to hunt as a unit. The Coyotes may split up, with some waiting in ambush while the others chase the prey toward them, or they may run in relays to tire their quarry—Coyotes, which are the best runners among the North American canids, typically cruise at 40–50 km/h.

Coyotes owe their modern-day success to their varied diet, early age of first breeding, high reproductive output and flexible living requirements. They consume carrion throughout the year, but they also feed on such diverse offerings as eggs, mammals, birds and berries. Their variable diet and nonspe-

RANGE: Coyotes are not found in the western third of Alaska, the tundra regions of northern Canada and the extreme southeastern U.S.; their range covers essentially the remainder of North America.

Total Length: 1.1–1.4 m
Shoulder Height: 58–70 cm
Tail Length: 30–40 cm
Weight: 10–22 kg

cific habitat choices allow them to adapt to just about any region of North America.

DESCRIPTION: Coyotes look like grey, buffy or reddish-grey, medium-sized dogs. The nose is pointed, and there is usually a grey patch between the eyes that contrasts with the rufous top of the snout. The bushy tail has a black tip. The underparts are light to whitish. When frightened, a Coyote runs with its tail tucked between its hindlegs.

HABITAT: Coyotes are found in all terrestrial habitats in North America, except the barren tundra of the far north and the humid southeastern forests. They have greatly expanded their range, mainly because of declining Grey Wolf numbers and forestry and agriculture that has brought about favourable changes in habitat for Coyotes.

DID YOU KNOW?

Coyotes can, and do, interbreed with domestic dogs. The "coydog" offspring often become nuisance animals, killing domestic livestock and poultry.

111

foreprint

walking trail

FOOD: Although primarily carnivorous, feeding on squirrels, mice, hares, birds, amphibians and reptiles, Coyotes will sometimes eat melons, berries and vegetation. Most farmers dislike Coyotes because they may take sheep, calves and pigs that are left exposed. They may even attack and consume dogs and cats.

DEN: The den is usually a burrow in a slope, frequently a Woodchuck hole that has been expanded to 30 cm in diameter and about 3 m in depth. Rarely, Coyotes have been known to den in an abandoned car, a hollow tree trunk or a dense brush pile.

YOUNG: A litter of 3 to 10 (usually 5 to 7) pups is born between late March and late May, after a gestation of about two months. The furry pups are blind at birth. Their eyes open after about 10 days, and they leave the den for the first time when they are three weeks old. Young Coyotes fight with each other and establish dominance and social position at just three to four weeks of age.

Grey Wolf

SIMILAR SPECIES: The **Grey Wolf** (p. 114) is generally larger, with a broader snout, much bigger feet and longer legs, and it carries its tail straight back when it runs. The **Red Fox** (p. 122) is generally smaller and much redder, and it has a white tail tip and black forelegs. The **Grey Fox** (p. 126) is smaller and has black spots on either side of the muzzle.

Grey Wolf

Canis lupus

For many North Americans, the Grey Wolf represents the apex of wilderness, symbolizing the pure, yet hostile, qualities of all that remains wild. Other people disparage this representation, characterizing wolves as blood-lusting enemies of domestic animals and the farmers who care for them.

Perhaps the persecution of wolves over the last few hundred years was not really against *Canis lupus*, but against the wolf-beast that lives only in the human imagination. The fear of wolves, without an understanding of their basic nature, resulted in tens of thousands of wolves killed in vengeance for crimes they did not commit. By studying and observing wolves, we demystify these beasts and learn acceptance and even admiration for these remarkable creatures.

Observations and behavioural studies of wolves indicate that the social structure of a wolf pack is extremely sophisticated. A pack behaves like a "super organism," cooperatively making it possible for more animals to survive. By hunting together, pack members can catch and subdue much larger prey than if they were acting alone.

A wolf pack can be described in terms of the alpha pair (top male and female), the subordinate adults, the outcasts and the pups and immature individuals.

Usually, only the alpha animals reproduce, while the other pack members help with bringing food to the pups and defending the group's territory. In most packs, the subordinate adults are non-breeding, although they might mate in spite of the rules, especially if the dominant pair is not paying attention or is otherwise occupied. If there is an outcast in the group, it is often picked on and usually gets just a shred of the good meals. The energetic pups of a wolf pack demand constant attention; they are always ready to pounce on their mother's head or tackle an unsuspecting sibling.

As the pups grow, they develop important skills that will aid them as adults. At the entrance to the den, they make their first attempts at hunting when they swat and bite at beetles and the occasional mouse or vole. These animals, however, do not react in quite the same way as larger prey does, and so the next step in the pups' apprenticeship as hunters is to watch their parents take down a big meal item such as a deer or Moose.

A wolf pack generally occupies a large territory—usually 260–780 km^2—so individual densities are extremely low. Moreover, because of widespread extermination efforts in North America in the last century, wolves are absent over

RANGE: Much reduced from historic times, the Grey Wolf's range currently covers most of Canada and Alaska, except the Prairies and southern parts of eastern Canada. It extends south into Minnesota and Wisconsin and along the Rocky Mountains into Idaho, Montana and Wyoming.

Total Length: 1.4–2 m
Shoulder Height: 65–100 cm
Tail Length: 35–50 cm
Weight: 25–80 kg

much of their historic range. Northern Ontario has very high and stable numbers of wolves, and if you visit the wilderness here you have an excellent chance of seeing or hearing wolves.

Recent research of wolves in the Algonquin area indicates that they may be a separate species from the Grey Wolf. If these findings are correct, the new species is called the Eastern Red Wolf (*C. lycaon*).

ALSO CALLED: Timber Wolf.

DESCRIPTION: A Grey Wolf resembles an over-sized, long-legged German Shepherd with extra-large paws. Although typically thought of as being a grizzled grey colour, a wolf's coat can range from coal black to creamy white. Black wolves are most common in dense forests; whitish wolves are characteristic of tundra regions. The bushy tail is carried straight behind the wolf when it runs. In social situations, the height of the tail generally relates to the social status of that individual.

HABITAT: Although wolves formerly occupied numerous habits throughout North America, they are now mostly restricted to forests, streamside woodlands and Arctic tundra.

FOOD: Grey Wolves primarily hunt cervids such as White-tailed Deer, Caribou and Moose. Although large mammals typically make up about 80

DID YOU KNOW?

Wolves are capable of many facial expressions, such as pursed lips, smile-like grins, upturned muzzles, wrinkled foreheads and angry, squinting eyes. Wolves even stick their tongues out at each other—a gesture of appeasement or submission.

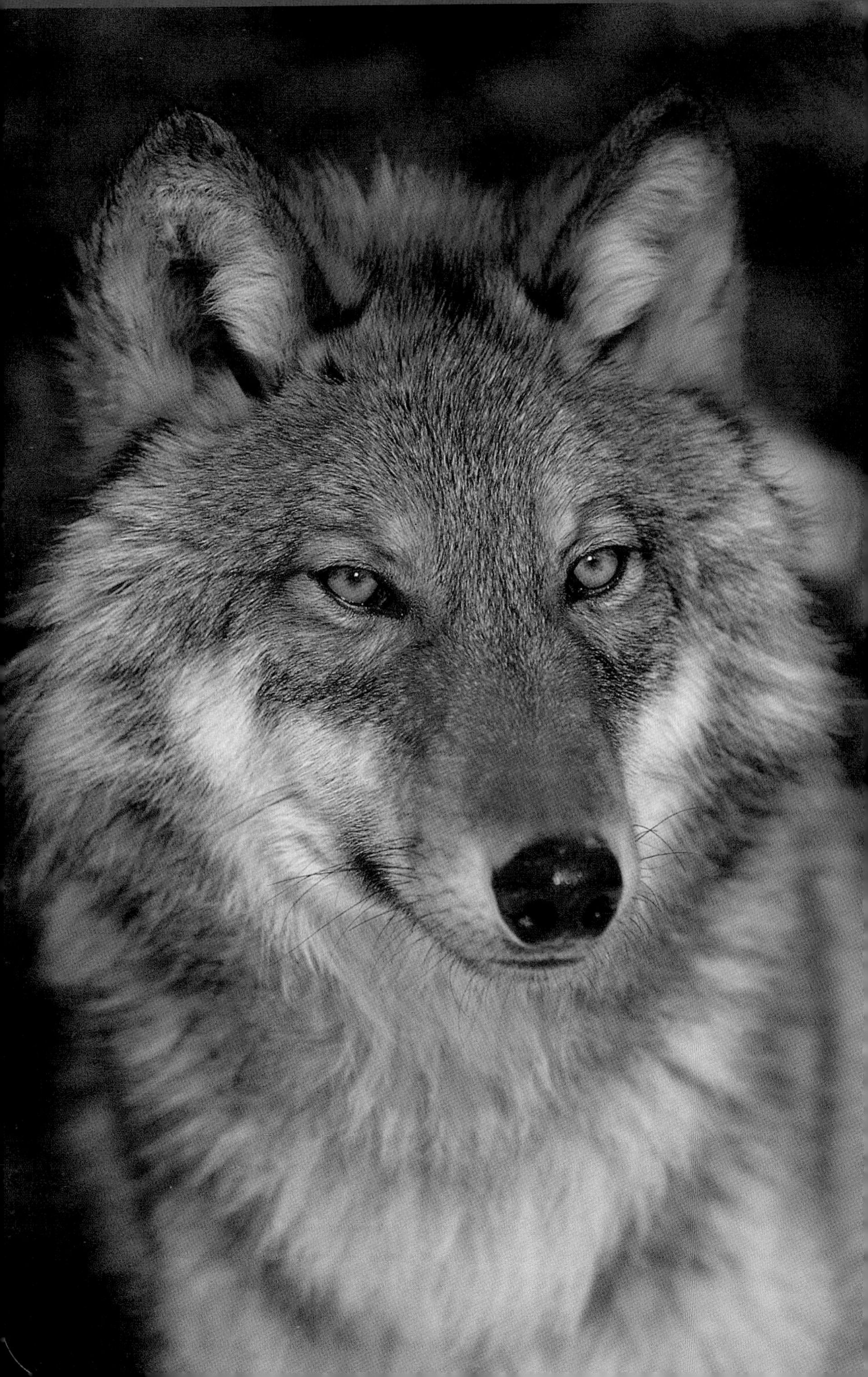

foreprint

walking trail

percent of the diet, wolves also prey on rabbits, mice, nestling birds and carrion when available.

DEN: Wolf dens are usually located on a rise of land near water. Most dens are bank burrows, and they are often made by enlarging the den of a fox or burrowing mammal. Sometimes a rock slide, hollow log or natural cave is used. Sand or soil scratched out of the entrance by the female is usually evident as a large mound. The burrow opening is generally about 60 cm across, and the burrow extends back 2–10 m to a dry natal chamber with a floor of packed soil. The beds from which adults can keep watch are generally found above the entrance.

YOUNG: A litter usually contains 5 to 7 pups (with extremes of 3 to 13), which may be of different colours. The newborn pups resemble domestic dogs in their development: their eyes open at 9 to 10 days, and they are weaned at 6 to 8 weeks. The pups are fed regurgitated food until they begin to accompany the pack on hunts. A wolf becomes sexually mature a couple of months before its third birthday, but the pack hierarchy largely determines the first incidence of mating.

Coyote

SIMILAR SPECIES: The **Coyote** (p. 110) is smaller, with a much more slender and pointed snout. A Coyote's tail is always tipped with black, and it is always held pointing downwards, as opposed to horizontally in a wolf. The **Red Fox** (p. 122) is much smaller and reddish in colour.

Arctic Fox
Alopex lagopus

Right or wrong, humans have a strong tendency to rank animal intelligence according to how many "human-like" behaviours the animal has. We talk about animals as "tool users" or "problem solvers," and generally assume that creatures demonstrating these traits have some level of intelligence.

The Arctic Fox exhibits a behaviour previously thought of as uniquely human. When we prepare too much for dinner, or we save summer foods for winter, we freeze them. Arctic Foxes, resourceful as they are, also freeze food for winter. In summer, food is plentiful, and they gluttonously feed on everything they can catch. When there is excess meat or carrion, the Arctic Fox digs a hole in the ground to the permafrost layer and puts the food underground, where the food stays frozen and stored until the meagre winter months.

Another behaviour of the Arctic Fox that has been observed is fishing—in a rather peculiar way. When simpler prey is scarce, an Arctic Fox may resort to approaching open water and tapping or stirring the surface with its paw. Fish are attracted to this movement at the surface, and when a fish comes within biting range the Arctic Fox quickly nabs it with its sharp, rapid-closing jaws. Cleverness has long been associated with foxes, and each species of fox has unique characteristics that epitomize the old saying, "Crafty as a fox."

The Arctic Fox typically occurs north of treeline, creatively surviving on Polar Bear or wolf kills, rodents, ground-nesting birds and their eggs and the occasional seal pup. It also seems to follow the migratory Caribou herds; even though Caribou are too large for this diminutive fox, wolves also follow the Caribou herds and leave many scraps on which the Arctic Fox dines. In the coldest, snowiest parts of winter, the Arctic Fox can be found a bit south of the tundra into the treeline. As soon as the snow begins to melt, however, the Arctic Fox returns to the northern tundra to feast on the bounty of the Arctic summer.

Because of its small size this fox falls prey to several other larger carnivores. Humans are probably the most relentless pursuers of this striking canine because of its luxurious thick coat. Its natural predators include Polar Bears (p. 106), Grey Wolves (p. 114) and Snowy Owls (which can take pups). Here in Ontario the range of the Red Fox (p. 122) overlaps with that of the Arctic Fox, and where overlap occurs the Red Fox is

RANGE: In North America, the circumpolar Arctic Fox ranges throughout the Arctic tundra, from western Alaska to Labrador. This fox has been seen as far north as 88° N on the polar ice, and there are a few records in Ontario from around the 49th parallel.

Total Length: 75–91 cm
Shoulder Height: 25–30 cm
Tail Length: 27–34 cm
Weight: 1.8–4.1 kg

summer coat

dominant. The Red Fox does not actively prey on the Arctic Fox, but it will harass and even kill its smaller cousin.

DESCRIPTION: This cat-sized fox is the only canid with distinct seasonal pelages. The summer coat is generally bluish-grey above and light or white below, with some white hairs on the head and shoulders, but most summer patterns are unique. The forelegs are brown and the tail is brownish above and lighter below, often with some long, white hairs. In winter, only the black nose and brownish-yellow eyes contrast with the long, white coat and bushy, white tail. On rare occasions, the winter pelage of an Arctic Fox is in a "blue" phase: the entire coat is blue-black to pearl grey. This fox is the only native member of the canid family that changes its coat colour seasonally. The ears are short and rounded, and the soles of the feet sport abundant hair between the toe pads.

HABITAT: Typically occurring in the treeless Arctic tundra or out on the polar ice, some Arctic Foxes may move south into the northern forests in winter.

FOOD: Arctic Foxes feed avidly on rodents, which they may dig up from under the snow. Their diet also includes birds, eggs, young hares, insects, carrion, seal pups and the scraps left by Polar Bears or wolves.

DID YOU KNOW?

In Siberia, the Arctic Fox has been known to move as much as 1200 km south during winter.

DEN: On the tundra, Arctic Foxes make their dens under rocks or in banks or hillsides, often with multiple entrances. The area around the entrance may be littered with feathers and bits of bone; sometimes rather large pieces of bones are carried to the den. In winter, these foxes may tunnel into snowbanks.

YOUNG: When there is an abundance of food, Arctic Foxes may mate as early as mid-February, but breeding is more typically delayed until late March or, if food is scarce, late April. In times of extreme hardship, Arctic Foxes might not mate at all. After a gestation of 1½ to 2 months, the vixen usually gives birth to six to nine helpless kits. Instances where up to 25 pups are found in a den may represent multiple litters. The young are weaned at two to four weeks. If food becomes scarce, the stronger pups may attack, kill and devour their siblings. The parents continue to feed their pups until mid-August. Arctic Foxes are sexually mature by their first birthday.

foreprint

walking trail

Grey Wolf

SIMILAR SPECIES: There is no other mammal that resembles an Arctic Fox in its white winter coat; a white **Grey Wolf** (p. 114) is much larger. The **Red Fox** (p. 122) is larger, has taller ears and is generally reddish. The **Grey Fox** (p. 126) is found much further south.

Red Fox
Vulpes vulpes

More than other native canids, the Red Fox has received some favourable representations in literature and modern culture. From *Aesop's Fables* to sexy epithets, the fox is often symbolized as a diabolically cunning, intelligent, attractive and noble animal. Many people favour having foxes nearby because of this species' skill at catching mice. In Old English, the fox has a well-deserved nickname, "reynard," from the French word *renard*, which refers to someone who is unconquerable owing to his or her cleverness. The fox's intelligence, undeniable comeliness and positive impact upon most farmlands have endeared it to many people who otherwise might not appreciate wildlife.

Foxes at work in their natural habitat embody playfulness, roguishness, stealth and drama. Young fox kits at their den wrestle and squabble in determined sibling rivalry. If its siblings are busy elsewhere, a young kit may amuse itself by challenging a plaything, such as a stick or piece of old bone, to a bout of aggressive mock combat. An adult out mousing will sneak up on its rustling prey in the grass and jump stiff-legged into the air, attempting to come down directly atop the unsuspecting rodent. If the fox misses, it stomps and flattens the grass with its forepaws, biting in the air to try to catch the fleeing mouse. Usually the fox wins, but, if not, it will slip away with stately composure as though the display of undignified abandon never occurred.

Oddly enough, Red Foxes exhibit both feline dexterity and a feline hunting style. Foxes hunt using an ambush style, or they creep along in a crouched position, ready to pounce on unsuspecting prey. Another un-dog-like characteristic is the large gland above the base of the tail, which gives off a strong musk somewhat resembling the smell of a skunk. This scent is what allows foxhounds to easily track their quarry. Foxes are territorial, and the males, like other members of the dog family, mark their territorial boundaries with urine.

The Red Fox is primarily nocturnal, and its keen senses of sight, hearing and smell enhance its elusive nature; at the first hint of an intruder, the fox will travel elsewhere. Winter may be the best time to see a fox: it is more likely to be active during the days, and its colour stands out when it is mousing in a snow-covered field. Red Foxes have adapted to human activity, and most of them live in farming communities and even in cities. In the northern wilderness, these mid-sized canids live on

RANGE: In North America, this holarctic species occurs throughout most of Canada and the U.S., except for the high Arctic, northwestern B.C. and much of the western U.S. The most widely distributed carnivore in the world, it also occurs in Europe, Asia, north Africa and as an introduced species in Australia.

Total Length: 90–110 cm
Shoulder Height: 38–41 cm
Tail Length: 35–45 cm
Weight: 3.6–6.8 kg

mice and carcasses in the shadow of the Grey Wolf (p. 114).

DESCRIPTION: This small, slender, dog-like fox has an exceptionally bushy, long tail. Its upperparts are usually a vivid reddish orange, with a white chest and belly, but there are many colour variations: a Coyote-coloured phase; the "cross fox," which has darker hairs along the back and across the shoulder blades; and the "silver fox," which is mostly black with silver-tipped hairs. In all colour phases, the tail has a white tip and the backs of the ears and fronts of the forelegs are black.

HABITAT: The Red Fox prefers open habitats interspersed with brushy shelter year-round. It avoids extensive areas of dense, coniferous forest with heavy snowfall.

FOOD: This opportunistic feeder usually stalks its prey and then pounces on it or captures it after a short rush. In winter, small rodents, rabbits and birds make up most of the diet, but dried berries are also eaten. In more moderate seasons, invertebrates, birds, eggs, fruits and

DID YOU KNOW?

The Red Fox's signature feature, its white-tipped, bushy tail, provides balance when the fox is running or jumping, and during cold weather a fox wraps its tail over its face.

foreprint

👣

👣

👣

👣

walking trail

berries supplement the basic small-mammal diet.

DEN: The Red Fox generally dens in a burrow, which the vixen either digs herself or, more usually, makes by expanding a Woodchuck hole. The den is sometimes located in a hollow log, in a brush pile or beneath an unoccupied building.

YOUNG: A litter of 1 to 10 kits is born in April or May after a gestation of about 7½ weeks. The kits weigh about 100 g at birth. Their eyes open after nine days, and they are weaned when they are one month old. The parents first bring the kits dead food and later crippled animals. The male may bring back to the den several voles or perhaps a hare and some mice at the end of a single hunting trip. After the kits learn to kill, the parents start taking them on hunts. The young disperse when they are three to four months old; they become sexually mature well before their first birthdays.

Coyote

SIMILAR SPECIES: The larger **Coyote** (p. 110) has a dark-tipped tail and does not have black forelegs. The **Grey Fox** (p. 126) is grizzled grey over the back and has a black-tipped tail, and the distinctive **Grey Wolf** (p. 114) can be more than twice the size. The **Arctic Fox** (p. 118) is smaller and lacks the reddish colour.

Grey Fox

Urocyon cinereoargenteus

Truly a crafty fox, the Grey Fox is known to elude predators by taking the most unexpected of turns—running up a tree. Unlike other canids, this fox seems comfortable in a tree, and it may climb into the branches to rest and sleep. There are even rare records of these foxes denning in natural tree cavities, and raising their litters as high as 6 m off the ground. It is the only North American member of the dog family that can climb.

The Grey Fox's great rarity and nocturnal habit mean it is less frequently seen than other foxes. Furthermore, it frequents treed areas, enhancing its elusive nature. This fox occurs in very low numbers in southernmost Ontario and between Thunder Bay and Rainy Lake. Your chances of seeing one are slim. Foxes, on the whole, are quite evasive. If you have an optimistic nature, you can turn your eyes skyward and scan the trees, especially those with thick, heavily forked trunks or leaning branches, for this species.

After the mating season, the male stays with the female and helps with raising the young. His primary role after the female gives birth is to bring her food, because she must remain with the young constantly for several days. The Grey Fox often caches its food, especially large kills that cannot be consumed at once. Small kills may be buried right near the den during the whelping season, partly to provide the female with ready food but also to stimulate the interest of the pups. Large cache sites are made either in heaped up vegetation or in holes dug into loose dirt.

Many of the fox populations in North America suffered great losses during the peak of the fur trade. To foxes, Coyotes and wolves, humans are the worst enemy. Fortunately for the Grey Fox, its pelt is of lower quality (to humans, that is—the fox certainly appreciates it) than the furs of other canids. The grizzled fur is very stiff, rather than soft and long like the winter coat of a Red Fox (p. 122). The Grey Fox has also suffered less persecution from farmers, because it is quite shy and rarely hunts domestic animals. Unlike other canids, the Grey Fox is not inclined toward chickens. It prefers to hunt the mice that abound around a henhouse, and its mousing ability is so good that it is even considered a welcome visitor to a barnyard.

DESCRIPTION: This handsome fox has an overall grizzled appearance because of the long, greyish fur over its back. Its undersides are reddish in colour, as are

RANGE: Grey Foxes have an extensive distribution in the U.S., and they occur in Canada along the southern borders of Ontario and Québec. They range through most of the eastern states and from Texas west to California and up the West Coast through most of Oregon. Some populations can be found in Colorado and Utah.

Total Length: 80–113 cm
Shoulder Height: 36–38 cm
Tail Length: 28–44 cm
Weight: 3.4–5.9 kg

the back of the head, throat, legs and feet. Sometimes the belly may be mostly white with only reddish highlights. The tail is grey or even black on top, with reddish undersides and a black tip. The ears are pointed and mainly grey in colour, with patches of red on the back side. A distinct black spot is present on either side of the muzzle.

HABITAT: The Grey Fox inhabits a variety of different environments, always near trees or groundcover. This fox prefers foraging in wooded areas rather than open environments.

FOOD: Grey Foxes are more omnivorous than other foxes. They consume a variety of small mammals, such as rabbits, rodents and birds, as well as large amounts of insects and other invertebrates. Late in summer, grasshoppers, crickets and other agricultural pests

DID YOU KNOW?

Although this fox is quite small, it can run very quickly over short distances. In one record, a Grey Fox topped 45 km/h.

foreprint

constitute much of the diet. A significant part of the diet is vegetable matter, such as fruits and grasses. Favourite items include apples and nuts.

DEN: Most commonly, this fox dens on a ridge or rocky slope or under brushy cover, but it may also den underground. If necessary, a fox digs a burrow itself, but it prefers to refurbish the abandoned burrow of another animal, such as a Woodchuck.

YOUNG: The Grey Foxes in Ontario may be non-breeding dispersing animals. In other areas, mating occurs in January or February, and after a gestation of about 53 days, one to seven young are born. They are born blind and almost hairless, and for the first several days they require constant care by the mother. After 12 days their eyes open, and they venture out of the den when they are 1½ or 2 months old. When they are four months old, they learn how to hunt and accompany their parents while foraging. By the fifth month, they have dispersed to start their own dens.

walking trail

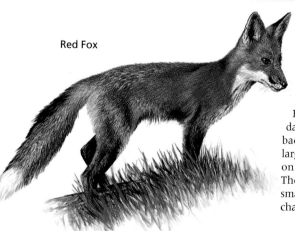

Red Fox

SIMILAR SPECIES: The **Red Fox** (p. 122) is much redder overall and has a white-tipped tail. A Red Fox "cross-fox" is usually darker over the shoulders and back. The **Coyote** (p. 110) is larger and lacks the black spots on either side of the muzzle. The **Arctic Fox** (p. 118) is smaller, lives further north and changes colour seasonally.

RODENTS

In terms of sheer numbers, rodents are the most successful group of mammals in Ontario. Because we usually associate rodents with rats and mice, the group's most notorious members, many people look on all rodents as filthy vermin. You must remember, however, that the much more endearing chipmunks, Beavers and squirrels are also rodents.

A rodent's best-known features are its upper and lower pairs of protruding incisor teeth, which continue to grow throughout the animal's life. These four teeth have pale yellow to burnt orange enamel only on their front surfaces; the soft dentine at the rear of each tooth is worn away by the action of gnawing, so that the teeth retain knife-sharp cutting edges. Most rodents are relatively small mammals, but beavers and porcupines can grow quite large.

Porcupine Family (Erethizontidae)

The stocky-bodied Porcupine has some of its hairs modified into sharp-pointed quills that it uses in defence. Its sharp, curved claws and the rough soles of its feet are adapted for climbing.

Porcupine

Jumping Mouse Family (Zapodidae)

Jumping mice are so called because they make long leaps when they are startled. The hindlegs are much longer than the forelegs, and the tail, which is longer than the combined length of the head and body, serves as a counterbalance during jumps. Jumping mice are almost completely nocturnal, and they hibernate.

Meadow
Jumping Mouse

Mouse Family (Muridae)

This diverse group of rodents is the largest and most successful mammal family in the world. Its members include the familiar rats and mice, as well as voles and lemmings. The representatives of this family in Ontario vary in size from the small Woodland Vole to the Muskrat.

Woodland Vole

Beaver Family (Castoridae)

The Beaver is one of two species worldwide in its family, and the only representative on our continent. It is the largest North American rodent. After humans, it is probably the animal with the biggest impact on the wilderness landscape of Ontario.

Beaver

Squirrel Family (Sciuridae)

This family, which includes chipmunks, tree squirrels, flying squirrels, marmots and ground squirrels, is considered the second-most structurally primitive group of rodents. All its members, except the flying squirrels, are active during the day, so they are seen more frequently than other rodents.

Least Chipmunk

131

Porcupine

Erethizon dorsatum

Although it lacks the charisma of large carnivores and ungulates, the Porcupine's claim to fame is its unsurpassed defensive mechanism. A Porcupine's formidable quills, numbering about 30,000, are actually stiff, modified hairs with shingle-like, overlapping barbs at their tips.

Contrary to popular belief, a Porcupine cannot throw its quills, but if it is attacked it will lower its head in a defensive posture and lash out with its tail. The loosely rooted quills detach easily, and they may be driven deeply into the attacker's flesh. The barbs swell and expand with blood, making the quills even harder to extract. Quill wounds may fester, or, depending on where the quills strike, they can blind an animal, prevent it from eating or even puncture a vital organ.

Porcupines are strictly vegetarian, and they are frequently found feeding in agricultural fields, willow-edged wetlands and forests. The tender bark of young branches seems to be a Porcupine delicacy, and though you wouldn't think it from their size, Porcupines can move far out on very thin branches with their deliberate climbing. Accomplished, if slow, climbers, Porcupines use their sharp, curved claws, the thick, bumpy soles of their feet, and the quills on the underside of the tail in climbing. Despite their agility in trees, they have been known to fall on occasion, breaking bones and even stabbing themselves with their quills. In winter, these large, stocky rodents often remain in individual trees and bushes for several days at a time. When they leave a foraging site, the gnawed, cream-coloured branches are clear evidence of their activity. In summer Porcupines can be seen in more open areas, foraging on herbaceous plants.

The Porcupine is primarily nocturnal in summer. It often rests by day in a hollow tree or log, in a burrow or in a treetop. Along the Niagara escarpment, it is common in rocky caves and crevices. In winter, it is not unusual to see a Porcupine active by day, either in an open field or in a forest. It often chews bones or fallen antlers for calcium, and the sound of a Porcupine's gnawing can sometimes be heard at a considerable distance.

Unfortunately for the Porcupine, its armament is no defence against vehicles—highway collisions are a major cause of Porcupine mortality—and most people only see Porcupines in the form of roadkill.

DESCRIPTION: This large, stout-bodied rodent has long, light-tipped guard hairs

RANGE: The Porcupine is widely distributed from Alaska across Canada to Pennsylvania and New England and south through most of the West into Mexico.

Total Length: 55–95 cm
Tail Length: 14–25 cm
Weight: 3.5–18 kg

surrounding the centre of the back, where abundant, long, thick quills crisscross one another in all directions. The young are mostly black, but the adults are variously tinged with yellow. The upper surface of the powerful, thick tail is amply supplied with dark-tipped, white to yellowish quills. The front claws are curved and sharp. The skin on the soles of the feet is covered with tooth-like projections. There may be grey patches on the cheeks and between the eyes.

HABITAT: Porcupines occupy a variety of habitats, including coniferous, mixed and deciduous forests and even croplands near wooded areas.

FOOD: Completely herbivorous, the Porcupine is like an arboreal counterpart of the Beaver (p. 154). It eats leaves, buds, twigs and especially young bark or the cambium layer of both broad-leaved and coniferous trees and shrubs. During spring and summer, it eats considerable amounts of herbaceous vegetation. In winter it subsists on bark. The Porcupine typically puts on weight during spring and summer and loses it during autumn and winter. It seems to have a profound fondness for salt, and it will chew wood handles, boots and other material that is salty from sweat or urine.

DEN: Porcupines prefer to den in rocky caves and crevices or beneath rocks, but they sometimes move into abandoned buildings, especially in winter. They are typically solitary animals, denning alone, but they may share a den during particularly cold weather. Sometimes a Porcupine will sleep in a treetop for weeks, avoiding any den site, while it works on eating the tree bark.

DID YOU KNOW?

The name "porcupine" comes from the Latin *porcospinus*, spiny pig, and underwent many variations—Shakespeare used the word "porpentine"—before its current spelling was established in the 17th century.

foreprint

walking trail

YOUNG: The Porcupine's impressive armament inspires many questions about how it manages to mate. The female does most of the courtship, although males may fight with one another, and she is apparently stimulated by having the male urinate on her. When she is sufficiently aroused, she relaxes her quills and raises her tail over her back so that mating can proceed. Following mating in November or December and a gestation of 6½ to 7 months—unusually long for a rodent—a single precocious porcupette is born in May or June. The young Porcupine is born with quills, but they are not dangerous to the mother—the baby is born headfirst in a placental sac with its soft quills lying flat against its body. The quills harden within about an hour of birth. Porcupines have erupted incisor teeth at birth, and although they may continue to nurse for up to four months, they begin eating green vegetation before they are one month old. Porcupines become sexually mature when they are 1½ to 2½ years old.

SIMILAR SPECIES: No other animal in Ontario closely resembles the Porcupine, but there is a small chance that, in a nocturnal sighting, the **Raccoon** (p. 88) could be mistaken for a Porcupine.

Raccoon

135

Meadow Jumping Mouse
Zapus hudsonius

Total Length: 19–22 cm

Tail Length: 11–14 cm

Weight: 15–25 g

On the rare occasions when these fascinating mice are encountered, their method of escape belies their true identities: startled from their sedgy homes, jumping mice hop away in a manner befitting a frog. Unfortunately, this rodent's speed and the abundance of hideouts prevent extended observations.

Jumping mice in northern Ontario spend up to nine months of the year dormant. These mice are true hibernators; they enter a state of deep sleep, their metabolism slows to the barest minimum, and they survive solely on the body's fat stores. Adults are underground by the end of August; only those few juveniles that are below their minimum hibernation weight are active until mid-September. In southern Ontario, they are active longer.

DESCRIPTION: The back is brownish with a dark dorsal stripe, the sides are yellowish, and the belly is whitish. Juveniles are much browner dorsally than adults. The long, naked tail is dark above and pale below, and there is no white tip. The hindfeet are greatly elongated.

HABITAT: Moist fields are preferred, but this jumping mouse also occurs in brush, marshes, brushy fields or even woods with thick vegetation.

FOOD: In spring, insects account for about half the diet. As the season progresses, the seeds of grasses and many forbs are eaten as they ripen. In summer and autumn, subterranean fungi form a significant portion of the diet.

DEN: The Meadow Jumping Mouse hibernates in a nest of finely shredded vegetation in a burrow or other protected site. Its summer nest is built on the ground or in small shrubs.

YOUNG: Breeding takes place within a week of the female's emergence from hibernation, typically in April or May, depending on latitude. She bears two to nine young after a 19-day gestation. A slow maturation follows: eyes open after two to five days, and nursing continues for a month. The young must achieve a certain minimum weight, or they will not survive their first hibernation. Some females may have a second litter after the first one leaves.

SIMILAR SPECIES: The **Woodland Jumping Mouse** (p. 137) has a distinct white-tipped tail.

RANGE: This jumping mouse is found from southern Alaska across most of southern Canada (except the Prairies) and south to northeastern Oklahoma in the West and northern Georgia in the East.

Woodland Jumping Mouse
Napaeozapus insignis

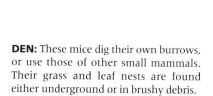

Total Length: 20–26 cm
Tail Length: 12–16 cm
Weight: 17–26 g

The remarkable Woodland Jumping Mouse is capable of making leaps up to 1.8 m in length or an astonishing 18 times the length of its own body. That's roughly equivalent to a human doing a 30-m dash in one leap. Even the babies seem born with a desire to jump. They require only about 12 days after birth to stand on their feet, albeit shakily, and immediately try out the hindlegs. By day 19, they regularly attempt jumps of about 2 or 3 cm, but they usually fall on their sides and have to wiggle to get back up. They are quick learners, however, and at day 28 they are easily leaping at least six times the length of their bodies.

DESCRIPTION: This jumping mouse is distinctly tricoloured. Its back is blackish to brownish, its sides are orange, and its undersides are white. The long tail is dark on top and light below, and it has a distinct white tip. The hindfeet are conspicuously large, and the forefeet are relatively small.

HABITAT: Woodland Jumping Mice are found in the dense foliage of moist forested regions—both coniferous and hardwood forests.

FOOD: The primary foods for these jumping mice include seeds, fungi, a variety of plant material and insects. These mice hibernate, but they do not store food. To survive winter, they put on a layer of fat before dormancy.

DEN: These mice dig their own burrows, or use those of other small mammals. Their grass and leaf nests are found either underground or in brushy debris.

YOUNG: After 29 days of gestation, the young are born in the nest in spring. Females have one litter per year of two to seven young. The young are altricial and require at least 34 days to gain the appearance of an adult.

SIMILAR SPECIES: The **Meadow Jumping Mouse** (p. 136) is slightly smaller and lacks the white-tipped tail. *Peromyscus* mice (pp. 142–44) have shorter tails and hindfeet.

RANGE: The Woodland Jumping Mouse occurs in southeastern Canada from Manitoba to Labrador and in the northeastern U.S. Southerly populations are found only in the Allegheny Mountains.

Norway Rat

Rattus norvegicus

While it is said that absence makes the heart grow fonder, it's a sure bet that no one misses rats. The cold, northern regions of Ontario are inhospitable to Norway Rats, but they are common in developed areas and farmland. Everywhere Norway Rats occur, they are subject to public scorn and intense pest control measures. These rats are not native to North America.

The geography and climate of much of Ontario help limit the spread of rats through the province. A rat is capable of dispersing 5–8 km in a summer, but if it is unable to find shelter in buildings or garbage dumps, winter temperatures of –18° C will prove fatal. The greatest influx of rats in the region comes courtesy of modern transportation. Rats hitchhiking on trucks and trains are of concern because they often get deposited in warm buildings in cities and towns throughout southern and central parts of Ontario.

Norway Rats were introduced to North America in about 1775, and they have established colonies in most cities and towns south of the boreal forest. These pests feed on a wide variety of stored grain, garbage and carrion, gnaw holes in walls, and contaminate stored hay with urine and feces. They have also been implicated in the transfer of diseases to both livestock and humans.

More than any other animal, the Norway Rat is viewed with disgust by most people. As one of the world's most studied and manipulated animals, however, much of our biomedical and psychological knowledge can be directly attributed to experiments involving these animals—a rather significant contribution for such a hated pest.

DESCRIPTION: The back is grizzled brown, reddish brown or black. The paler belly is greyish to yellowish white. The long, round, tapered tail is darker above and lighter below. It is sparsely haired and scaly. The prominent ears are covered with short, fine hairs. Occasionally, someone releases an albino, white or piebald Norway Rat that had been kept in captivity.

HABITAT: Norway Rats nearly always live in proximity to human habitation, and they can only live away from human structures in the warmest parts of the province. Where they are found away from humans, they prefer thickly vegetated regions with abundant cover. Abandoned buildings in the wilderness are more frequently occupied by native rodents than by Norway Rats.

RANGE: The Norway Rat is concentrated in cities, towns and farms throughout coastal North America, southern Canada (except Alberta) and most of the U.S.

Total Length: 33–46 cm
Tail Length: 12–22 cm
Weight: 200–480 g

FOOD: This rat eats a wide variety of grains, insects, garbage and carrion. It may even kill the young of chickens, ducks and other small animals. Green legume fruits are also popular items, and some shoots and grasses are consumed.

DEN: A cavity scratched beneath a fallen board or a space beneath an abandoned building may hold a bulky nest of grasses, leaves and often paper or chewed rags. Although Norway Rats are able to, they seldom dig long burrows.

YOUNG: After a gestation of 21 to 22 days, 6 to 22 pink, blind babies are born. The eyes open after 10 days. The young are sexually mature in about three months. In Ontario, Norway Rats seem to breed mainly in the warmer months of the year, but in some large cities they may breed year-round.

SIMILAR SPECIES: The **Muskrat** (p. 150) is larger and has a laterally compressed tail.

DID YOU KNOW?

Some historians attribute the end of the Black Death epidemics in Europe to the southward invasion of the Norway Rat and its displacement of the less aggressive Black Rat, which was much more apt to inhabit human homes.

House Mouse
Mus musculus

Thanks to its habit of catching rides with humans, first aboard ships and now in train cars, trucks and containers, the House Mouse is found in most countries of the world. In fact, the House Mouse's dispersal closely mirrors the agricultural development of our species. As humans began growing crops on the great sweeping plains of middle Asia, this mouse, native to that region, began profiting from our storage of surplus grains and our concurrent switch from a nomadic to a relatively sedentary lifestyle. It is not native to North America.

Within the small span of a few hundred human generations, farmed grains began to find their way into Europe and Africa for trade. Along with these grain shipments, stowaway House Mice were spread to every corner of the globe. Even in the harsh parts of northern Ontario, the House Mouse is found wherever humans provide it free room and board. Unlike many of the introduced animals in the region, however, this species seems to have had a minimal negative impact on native animal populations.

House Mice are known to most people who have spent some time on farms or in warehouses, university labs and disorderly places. The white mice commonly used as laboratory animals are an albino strain of this species.

DESCRIPTION: The back is yellowish brown, grey or nearly black, the sides may have a slight yellow wash, and the underparts are light grey. The nose is pointed and surrounded by abundant whiskers. There are large, almost hairless ears above the protruding, black eyes. The long, tapered tail is hairless, grey and slightly lighter below than above. The brownish feet tend to be whitish at the tips.

HABITAT: This introduced mouse inhabits homes, outbuildings, barns, granaries, haystacks and trash piles. It cannot tolerate temperatures below −10° C around its nest and seems to be unable to survive winters in the northern forests without access to heated buildings or haystacks. In summer, it may disperse slightly more than 3 km from its winter refuge into fields and grasslands, only to succumb the following winter. Deer Mice are far more likely to invade wilderness cabins than House Mice.

FOOD: Seeds, stems and leaves constitute the bulk of the diet, but insects, carrion and human food, including meat and milk, are eagerly consumed.

RANGE: The House Mouse is widespread in North America, inhabiting nearly every city, hamlet or farm from the Atlantic to the Pacific and north to the tundra.

Total Length: 13–20 cm
Tail Length: 6–10 cm
Weight: 14–25 g

DEN: The nest is constructed of shredded paper and rags, sometimes fur and vegetation combined into a 10-cm ball beneath a board, inside a wall, in a pile of rags or in a haystack. It may occur at any level in a building. House Mice sometimes dig short tunnels, but they generally do not use them as nest sites.

YOUNG: If abundant resources are available, as in a haystack, breeding may occur throughout the year, but populations away from human habitations seem to breed only during the warmer months. Gestation is usually three weeks, but it may be extended to one month if the female is lactating when she conceives. The litter usually contains four to eight helpless, pink, jellybean-shaped young. Their fur begins to grow in 2 to 3 days, the eyes open at 12 to 15 days, and they are weaned at 16 to 17 days. At six to eight weeks, the young become sexually mature.

SIMILAR SPECIES: The **Deer Mouse** (p. 142) and the **White-footed Mouse** (p. 144) have bright white undersides, dark dorsal stripes and distinctively bicoloured tails.

DID YOU KNOW?

The word "mouse" probably derives from the Sanskrit *mus*—also the source, via Latin, of the genus name—which itself came from *musha*, meaning "thief."

Deer Mouse
Peromyscus maniculatus

Upon first seeing a Deer Mouse, many people are surprised by its cute appearance. The large, protruding, coal black eyes give it a justifiably inquisitive look, while its dainty nose and long whiskers continually twitch, sensing the changing odours in the wind.

Wherever there is groundcover, from thick grass to deadfall, Deer Mice scurry about with great liveliness. These small mice are omnipresent over much of their range, and they may well be the most numerous mammal in Ontario. When you walk through forested wilderness areas, they are in your company, even if their presence remains hidden.

Deer Mice most frequently forage along the ground, but regularly climb into trees and shrubs to reach food. During winter, Deer Mice are the most common of the small rodents to travel above the snow. In doing so, however, they are vulnerable to nighttime predators. The tiny skulls of these rodents are among the most common remains in the regurgitated pellets of owls, a testament to their ubiquity and importance in the food web.

The Deer Mouse, which is named for the similarity of its colouring to that of the White-tailed Deer (p. 30), commonly occupies farm buildings, garages and storage sheds, often alongside the House Mouse (p. 140). Three subspecies of the Deer Mouse are known to occur in Ontario, each living in a slightly different habitat.

DESCRIPTION: Every Deer Mouse has protruding black eyes, large ears, a pointed nose, long whiskers and a sharply bicoloured tail, with a dark top and light underside. In contrast to these constant characteristics, the colour of the adult's upperparts is quite variable: yellowish buff, tawny brown, greyish brown or blackish brown. The upperparts, however, are always set off sharply from the bright white undersides and feet. A juvenile has uniformly grey upperparts.

HABITAT: These ubiquitous mice occupy a variety of habitats, including grasslands, mossy depressions, brushy areas, tundra and heavily wooded regions. Another habitat these little mice greatly favour is the human building—our warm, food-laden homes are palatial residences to Deer Mice.

FOOD: Deer Mice use their internal cheek pouches to transport large quantities of seeds or fruits from grasses,

RANGE: The Deer Mouse is the most widespread mouse in North America. Its range extends from Labrador almost to Alaska and south through most of North America to south-central Mexico.

Total Length: 14–21 cm
Tail Length: 5–10 cm
Weight: 18–35 g

chokecherries, buckwheat and other plants to their burrows for later consumption. They also eat insects, nestling birds and eggs.

DEN: As the habitat of this mouse changes, so does its den type: in meadows it nests in a small burrow or makes a grassy nest on raised ground; in wooded areas it makes a nest in a hollow log or under debris. Nests can also be made in rock crevices, and certainly in human structures. These mice are not meticulous about keeping their nest clean; every few weeks they abandon their nest and start fresh.

YOUNG: Breeding takes place between March and October, and gestation lasts for three to four weeks. Females have multiple litters in one season. The helpless young number one to nine (usually four or five) and weigh about 2 g at birth. They open their eyes between days 12 and 17, and about

four days after that they venture out of the nest. At three to five weeks the young are completely weaned and are soon on their own. A female is sexually mature in about 35 days, and a male in about 45 days.

SIMILAR SPECIES: In the field, the **White-footed Mouse** (p. 144) is difficult to distinguish, although it has a smaller range in the province. The **House Mouse** (p. 140) lacks the distinct bright white belly. **Jumping mice** (pp. 136–37) have much longer tails.

DID YOU KNOW?

Adult Deer Mice displaced more than a kilometre from where they were trapped were generally able to return to their home burrows within a day. Many other animals can also perform this feat, but no one fully understands the mechanism at work.

White-footed Mouse
Peromyscus leucopus

Total Length: 15–20 cm

Tail Length: 6–9.5 cm

Weight: 15–25 g

The two *Peromyscus* mice in Ontario are nearly impossible to tell apart without a specimen in hand. When measurements are obtained, this mouse has a slightly shorter tail than the Deer Mouse (p. 142). Nevertheless, confusion between the two species is common, even with a specimen and a mammal key.

This species demonstrates a strong swimming ability, and dispersing individuals are frequently found colonizing islands of lakes. Next to the Deer Mouse, the White-footed Mouse is able to live in a wider variety of habitats then most other *Peromyscus* species. The only major habitat component for this excellent climber is some form of canopy or shrub cover.

DESCRIPTION: This species has much variation, but most are pale to dark reddish brown above and white below. The tail is similarly bicoloured—brownish or greyish above and white below. The large ears extend well above the fur on the head, and the beady, black eyes protrude. The feet and heels are white.

HABITAT: The White-footed Mouse lives in a variety of habitats, including open woodlands, riparian areas, shrubby areas and some agricultural lands bordering wooded areas. It is not considered detrimental to agriculture, but it is known to occupy buildings as often as Deer Mice.

FOOD: These mice have a diverse, omnivorous diet that changes seasonally. Insects, seeds, assorted vegetation, berries and nuts are the primary foods.

DEN: Nests are located in logs, tree cavities, buildings or other sheltered areas. The nest is a sphere of grass and fine dry vegetation about 10 cm in diameter. These mice may refurbish and "cap" an abandoned bird's nest.

YOUNG: The breeding season extends from March to October, and females have multiple litters in one season. Following a gestation of at least 22 days, four to six young are born. The pink, naked, blind young weigh under 2 g at birth and are weaned about three to four weeks later.

SIMILAR SPECIES: The **Deer Mouse** (p. 142) is difficult to distinguish in the field; a specimen and key is needed. The **House Mouse** (p. 140) lacks the distinct bright white belly.

RANGE: The White-footed Mouse is found in the eastern U.S., north to southern Ontario, southwest as far as Arizona and northwest as far as extreme southern Saskatchewan.

Southern Red-backed Vole
Clethrionomys gapperi

Total Length: 12–16 cm

Tail Length: 3–6 cm

Weight: 12–43 g

FOOD: Green vegetation, grasses, berries, lichens, seeds and fungi form the bulk of the diet.

This attractive little vole, which is active both day and night, can be heard rustling in the leaf litter of forests throughout Ontario. It is almost never seen, however, because it scurries along on its short legs through almost invisible runways on the forest floor.

The Southern Red-backed Vole is a classic example of a subnivean wanderer, a small mammal that lives out cold winters between the snowpack and the frozen ground. The snow's insulating qualities create a layer at ground level within which the temperature is nearly constant. This vole does not cache food; instead, it forages widely under the snow for vegetation or any other digestible foods.

DESCRIPTION: The reddish dorsal stripe makes this animal one of the easiest voles to recognize. On rare occasions, the dorsal stripe is a rich brownish black or even slate brown. The sides are greyish buff, and the undersides and feet are greyish white. Compared with most voles, the black eyes seem small and the nose looks slightly more pointed. The short tail is slender and scantily haired, and it is grey below and brown above. The ears are rounded and project somewhat above the thick fur.

HABITAT: This vole is found in a variety of habitats, including damp coniferous forests, bogs, swampy land and sometimes drier forests.

DEN: Summer nests, made in shallow burrows, rotten logs or rock crevices, are lined with fine materials, such as dry grass, moss and lichens. Winter nests are subnivean: above the ground but below the snow.

YOUNG: Mating occurs between April and October. Following a gestation of about 20 days, two to eight (usually four to seven) pink, helpless young are born. They nurse almost continuously, and their growth is rapid. By two weeks they are well-furred and have opened their eyes. Once the young are weaned, they are no longer permitted in the vicinity of the nest. This vole reaches sexual maturity at two to three months.

SIMILAR SPECIES: The **Woodland Vole** (p. 149), rare in southern Ontario, has a shorter tail and is usually reddish on its sides and brownish on its back. Other *Microtus* voles are not as red. The **Eastern Heather Vole** (p. 146) has a shorter tail and is grizzled brown above.

RANGE: This vole is widespread across most of the southern half of Canada and south through the Cascade and Rocky mountains, as far as northern New Mexico and through the Appalachians to North Carolina.

Eastern Heather Vole
Phenacomys ungava

Total Length: 12–16 cm
Tail Length: 3–4 cm
Weight: 25–40 g

S tories abound among naturalists of how gentle this little vole is when captured. Among the variety of explanations for this tranquillity, the most unusual is that diet may play an important role in temperament. These voles feed heavily on willow bark, and certain compounds in willow bark and leaves are known to have calming and even analgesic properties. (Willows are the source of salicylic acid, the natural precursor to aspirin.) Other animals, such as the ptarmigan, also feed on willow bark, and they, too, have gentle demeanours.

DESCRIPTION: This vole is greyish brown over its back, with light grey to steel grey undersides. The tops of its feet are silvery grey. Generally, there are a few orange hairs at the base of the ears. The eyes are small, the ears are short and rounded, and the face is yellowish.

RANGE: The range of the Eastern Heather Vole closely corresponds to the range of the boreal forest from the Yukon down the eastern edge of the Rockies and across to Labrador.

The fur is long and silky, and the tail is white beneath and grey above.

HABITAT: These voles occur in a variety of habitats, but primarily open coniferous forests and shrubby areas on the edges of forests. Birch and willow thickets often attract this species.

FOOD: The green foliage of shrubs and forbs forms the bulk of the diet in summer. The bark and buds of shrubs form the bulk of the winter diet.

DEN: The nest is made of heather twigs and lichens gathered into a 15-cm ball and located in a sheltered place. The natal den is in a burrow in a rocky area, under a log, beneath a root or stump or at the base of a shrub.

YOUNG: Females mate from May to August. After a gestation of 19 to 24 days, 2 to 8 altricial young are born. Their eyes open at 14 days, whereupon they are weaned and begin eating vegetation. Females can mate as young as six weeks. Males mate after their first winter.

SIMILAR SPECIES: The **Southern Red-backed Vole** (p. 145) has a distinct reddish dorsal stripe and a longer tail. The larger **Meadow Vole** (p. 147) has blackish fur on the tops of its feet.

Meadow Vole
Microtus pennsylvanicus

Total Length: 13–19 cm
Tail Length: 3–5 cm
Weight: 18–64 g

The Meadow Vole is well adapted to the winters of Ontario. It is subnivean in winter, meaning it remains active beneath the snow but above the ground. When the snows recede from the land every spring, an elaborate network of Meadow Vole activity is exposed to the world. Runways, chambers and nests, previously insulated from winter's cold by deep snows, await the growth of spring vegetation to conceal them once again. These tunnels often lead to logs, boards or shrubs, where a vole can find additional shelter.

These voles are an important prey species—many of them die in their first months, and very few live as long as a year. With two main reproductive cycles a year, it is unlikely that many voles get to experience all seasons.

DESCRIPTION: The body varies from brown to blackish above and grey below. The protruding eyes are small and black. The rounded ears are mostly hidden in the long fur of the rounded head. The tops of the feet are blackish brown. The tail is about twice as long as the hindfoot.

HABITAT: The Meadow Vole can be found in a variety of habitats, provided grasses are present. Grasslands, pastures, marshy areas, open woodlands and tundra are all potential homes for this vole.

FOOD: The green parts of sedges, grasses and some forbs make up the bulk of the spring and summer diet. In winter, large amounts of seeds, some bark and insects are eaten. Other foods include grains, roots and bulbs.

DEN: The summer nest is made in a shallow burrow and lined with fine materials, such as dry grass, moss and lichens. The winter nest is subnivean: above the ground but below the snow.

YOUNG: Spring mating occurs between late March and the end of April. Gestation is about 20 days, and the average litter size is four to eight young. From birth, the helpless young nurse almost constantly to support their rapid growth. Their eyes open in 9 to 12 days, and they are weaned at 12 to 13 days. At least one more litter is born, usually in autumn.

SIMILAR SPECIES: The **Eastern Heather Vole** (p. 146) is smaller and has white feet. The **Rock Vole** (p. 148) has a longer tail and yellowish nose, and the **Woodland Vole** (p. 149) has reddish sides.

RANGE: This vole, the most widespread in North America, ranges from central Alaska to Labrador and south to northern Arizona and New Mexico in the West and Georgia in the East.

Rock Vole

Microtus chrotorrhinus

Total Length: 14–19 cm
Tail Length: 4.2–6.4 cm
Weight: 30–48 g

A common—and usually true—assertion is that voles make up a large percentage of the mammal population in nearly any natural community. Alas, the Rock Vole seems to have difficulty achieving a ubiquitous status. This vole has been identified in Pleistocene deposits, and in each case of fossil documentation it was the least common microtine rodent. Even today, when biologists trap microtines, the Rock Vole is only present in one-third the numbers of the other voles. This low proportion of individuals seems to be unchanging over time, even in the most favourable habitats. No one knows what causes this natural check on Rock Vole populations.

ALSO CALLED: Yellow-nosed Vole.

RANGE: Rock Voles are found from south-central Ontario and extreme northeastern Minnesota to Labrador and south to New York and parts of New England. An isolated population occurs in the Smoky Mountains.

DESCRIPTION: This medium-sized vole is tawny or brown above and grey below. The relatively long tail is similarly coloured. A helpful field mark to identify this vole is the colour of the nose—usually it is yellowish, orangish or slightly pink in appearance.

HABITAT: The Rock Vole inhabits rocky areas in moist woodlands. This vole requires water nearby its home.

FOOD: Rock Voles feed heavily on bunchberries, other fruit, grasses, leaves and fungi. Insects and larvae are also eaten.

DEN: These voles form extensive surface runways and subterranean burrows. Nests are either inside a shallow burrow or in a rock crevice. Separate burrow chambers or designated surface areas are used as latrines.

YOUNG: Mating occurs several times between early spring and late autumn. Females may have two or three litters per year, with one to seven young per litter. The young are altricial at birth.

SIMILAR SPECIES: The **Meadow Vole** (p. 147) has a shorter tail, and it and all others in Ontario lack the yellowish-orange nose.

Woodland Vole
Microtus pinetorum

Total Length: 11–15 cm
Tail Length: 1.2–2.9 cm
Weight: 19–39 g

Although the Woodland Vole may be the most abundant vole species in many regions south of Ontario, it is probably one of the rarest voles in our province. It can only be found in a few isolated areas in the Carolinian Forest region. Because of its rare distribution, and its tendency for a somewhat subterranean existence, you are not likely to see one of these handsome voles. If you see one, it will look very similar to the much more common Southern Red-backed Vole (p. 145). The two voles are not likely to have any range overlap, so identification should be easy. Despite being from a different genus (*Clethrionomys*), the red-backed vole looks similar to the Woodland Vole, except for its longer tail.

DESCRIPTION: This vole is quite small and has a very short tail. It is faintly tri-coloured, with a brownish back, reddish sides and greyish belly. Its short ears and small eyes are largely hidden in the fur. It has long whiskers and a rounded head, and the snout is blunt. The body appears cylindrical, with short limbs. Some individuals are greyish overall.

HABITAT: This vole is found mainly in wooded areas and thick, shrubby areas, especially where sandy soils are available for burrowing. Sometimes individuals are found in heavy grass areas.

FOOD: The Woodland Vole feeds mainly on forbs, grasses, roots, seeds, fruit, bark and sometimes insects and fungi. The choice of food varies depending on seasonal availability. This vole stores large amounts of food in underground burrows.

DEN: This vole digs burrows, and its nest chamber is located in the burrow and typically has multiple exits. The nest is lined with dried grasses, leaves and roots.

YOUNG: The Woodland Vole may have several litters (one to four) in a year, and reproduction usually takes place between spring and autumn. Gestation is about 21 days, and the average litter size is only two young.

SIMILAR SPECIES: The **Southern Red-backed Vole** (p. 145) has a longer tail and is reddish on the back rather than on the sides. The **Meadow Vole** (p. 147) lacks the reddish colour.

RANGE: The Woodland Vole is found throughout the eastern U.S. and just into Ontario in the Carolinian Forest.

Muskrat

Ondatra zibethicus

Although some Muskrats live winters in ice-free climates, the ones in Ontario are restricted to living beneath the ice of their pond and in their burrows. When the snow and ice melt, Muskrats might be seen out on dry land. In early spring many first-year animals, now sexually mature, venture from their birth ponds to establish their own territories. These dispersing Muskrats are commonly seen travelling over land. It is a tragic requirement for many—their numbers are all too easily tallied on May roadkill surveys.

The Muskrat is not a "mini-Beaver," nor is it a close relative of that large rodent; rather, it is a highly specialized aquatic vole that shares many features with the Beaver as a result of their similar environments. Like a Beaver (see p. 154), a Muskrat can close its lips behind its large, orange incisors, which allows it to chew underwater without getting water or mud in its mouth. Its eyes are placed high on its head, and a Muskrat can often be seen swimming with its head and sometimes its tail above water. The Muskrat dives with ease; it can remain submerged for over 15 minutes and can swim the length of a football field before surfacing.

Muskrats lead busy lives. They are continually gnawing cattails and bulrushes, whether they are eating the tender shoots or gathering the coarse vegetation for home building. Muskrat homes rise above shallow waters throughout the region, and they are of tremendous importance not only to these aquatic rodents, but also to geese and ducks, which make use of Muskrat homes as nesting platforms.

Both sexes have perineal scent glands that enlarge and produce a distinctly musk-like discharge during the breeding season. Although this scent is by no means unique to the Muskrat, its potency is sufficiently notable to have influenced this animal's common name. An earlier name for this species was "musquash," from the Abnaki name *moskwas,* but through the association with musk the name changed to "muskrat."

DESCRIPTION: The coat generally consists of long, shiny, tawny to nearly black guard hairs overlying a brownish-grey undercoat. The flanks and sides are lighter than the back. The underparts are grey, with some tawny guard hairs. The long tail is black, nearly hairless, scaly and laterally compressed with a dorsal and ventral keel. The legs are short. The hindfeet are large and partially webbed and have an outer fringe

RANGE: This wide-ranging rodent occurs from the southern limit of the Arctic tundra across nearly all of Canada and the lower 48 states, except most of Florida, Texas and California.

Total Length: 46–61 cm
Tail Length: 20–28 cm
Weight: 0.8–1.6 kg

of stiff hairs. The tops of the feet are covered with short, dark grey hair. The claws are long and strong.

HABITAT: Muskrats occupy sloughs, lakes, ponds, marshes and streams that have cattails, rushes and open water. Muskrats often occupy Beaver ponds; the two coexist harmoniously.

FOOD: The summer diet includes a variety of emergent herbaceous plants. Cattails, rushes, sedges, irises, water lilies and pondweeds are staples, but a few frogs, turtles, clams, snails, crayfish and an occasional fish may be eaten. In winter, Muskrats feed on submerged vegetation.

DEN: Muskrat houses are built entirely of herbaceous vegetation, without the branches or mud of Beaver lodges. The dome-shaped piles of cattails and rushes have an underwater entrance. Muskrats may also dig bank burrows, which are 5–15 m long and have entrances that are below the usual water level.

YOUNG: Breeding takes place between March and September. Each female produces two or sometimes three litters a year. Gestation lasts 25 to 30 days, after which six to seven young are born. The eyes open at 14 to 16 days, the young are weaned at three to four weeks, and they are independent at one month. Both males and females are sexually mature the spring after their birth.

SIMILAR SPECIES: The **Beaver** (p. 154) is larger and has a broad, flat tail, and typically only its head is visible above water when it swims. The **Norway Rat** (p. 138) is smaller and has a cylindrical tail. The similar-sized **Woodchuck** (p. 162) has a furred tail and is not associated with water.

DID YOU KNOW?

Muskrats are highly regarded by native peoples. In one story, it was Muskrat who brought some mud from the bottom of the flooded world to the water's surface. This mud was spread over Turtle's back, thus creating all the dry land we now know.

Southern Bog Lemming
Synaptomys cooperi

Total Length: 12–15 cm
Tail Length: 1.3–2.4 cm
Weight: 21–50 g

The Southern Bog Lemming is the most southerly of all lemmings. While most lemmings inhabit Arctic or near Arctic regions, this lemming prefers grassy meadows and open forests in southern Canada and the central and eastern U.S. Unlike its northern relative, the Northern Bog Lemming (p. 153), this lemming does not inhabit bog areas. In suitable habitat, this vole's presence is indicated by well-used runways and neatly clipped piles of grass at intervals along the paths. As well, little green fecal pellets hint at recent activity. Where the range of this lemming and the Meadow Vole (p. 147) overlap, the vole easily outcompetes the lemming. This bog lemming does not appear to have a high density anywhere in Ontario.

DESCRIPTION: This vole is brownish above and silvery or greyish below. It has a very short tail that is faintly bicoloured—brown above and greyish below. In summer, the nape of the neck and shoulders may be rich brown. The strong, curved claws aid in digging the elaborate winter runways.

HABITAT: Open forests, grassy meadows and shrub or sedge areas are suitable habitats for these lemmings.

FOOD: Grasses, clover and sedges compose the majority of the diet, although fungi, algae, roots, bark and mosses are occasionally eaten. In times of scarce vegetation, any emergent plant is eaten down to the surface.

DEN: This lemming digs its own burrows, and it may use the burrows of other small mammals. The nest can be above or below ground. An above ground nest is usually located in a sheltered area and looks like a spherical grassy clump. A nest below ground is usually less than 15 cm below the surface in a chamber off the main burrow.

YOUNG: Breeding occurs from spring through autumn and sometimes in winter in southern populations. Gestation is about 25 days, after which one to eight young are born. The young are weaned at 16 to 21 days, and they are probably sexually mature soon thereafter.

RANGE: The Southern Bog Lemming can be found from southeastern Manitoba to Newfoundland and south to Kansas in the West and Virginia in the East.

SIMILAR SPECIES: The **Northern Bog Lemming** (p. 153) has a reddish spot at the base of the ears. Most voles in Ontario have tails longer than 2.4 cm.

Northern Bog Lemming
Synaptomys borealis

Total Length: 11–14 cm

Tail Length: 1.7–2.7 cm

Weight: 23–35 g

The Northern Bog Lemming is found in northern Ontario. Its favoured habitat is cool sphagnum bogs, but black spruce forests and tundra sedge meadows can also host populations. This bog lemming can be found at sea level in suitable habitat along the Hudson Bay coast.

Although these animals are rarely seen, their workings are easy enough to identify. The mossy runways are frequently marked by evenly clipped grasses stacked in neat piles, like harvested trees awaiting logging trucks along haulroads. Another sign of their recent activity is their greenish fecal pellets in little piles along the runways.

DESCRIPTION: The ears of this stout lemming scarcely project above the fur of the head. The whole body is covered in thick fur. Although there are various colour phases, the sides and back are usually chestnut or dark brown, and the underparts are usually greyish. A little patch of tawny or reddish hair is just behind the ears. The claws are strong and curved, and those on the middle two front toes become greatly enlarged in winter to aid digging in frozen conditions.

HABITAT: This lemming thrives in wet tundra conditions, such as tundra bogs, meadows and even spruce woods.

FOOD: The diet is primarily composed of grasses, sedges and similar plants. If this vegetation is scarce, other emergent plants are eaten.

DEN: In summer, nests are located in tunnels about 15 cm underground. The nests are made of dry grass and fur, and there are nearby chambers for wastes. In winter, the nests are located aboveground, but under the snow.

YOUNG: Little is known about the reproduction of the Northern Bog Lemming, but it is thought to breed between spring and autumn, with a gestation of about three weeks. The litter contains two to six helpless young. Growth is rapid: they are furred by one week, weaned by three weeks, and leave to start their own families soon thereafter.

SIMILAR SPECIES: The **Southern Bog Lemming** (p. 152) lacks the reddish spot at the base of the ears. Most voles in Ontario have longer tails.

RANGE: The Northern Bog Lemming ranges across most of Alaska and Canada south of the Arctic tundra. It occurs as far south as northern parts of Washington, Idaho and western Montana.

Beaver

Castor canadensis

The Beaver is truly a great North American mammal. Its highly valued pelt motivated the earliest explorers to discover the riches of the wilderness, and, even today, the Beaver serves as an international symbol for wild places. Quite surprisingly to many Canadians, foreign tourists often hold great hopes of seeing these aquatic specialists during their visits. Fortunately, the Beaver can be found in wet areas throughout Ontario (except in winter), where its engineering marvels can be studied in awe-inspiring detail.

One of the few mammals that significantly alters its habitat to suit its needs, the Beaver often sets back ecological succession and brings about changes in vegetation and animal life. Nothing seems to bother a Beaver like the sound of running water, and this busy rodent builds dams of branches, mud and vegetation to slow the flow of water. The deep pools that the Beaver's dams create allow it to remain active beneath the ice in winter, at a cost of vast amounts of labour—a single Beaver may cut down hundreds of trees each year to ensure its survival.

Beavers live in groups that generally consist of a pair of mated adults, their yearlings and a set of young kits. This family group usually occupies a tightly monitored habitat that consists of several dams, terrestrial runways and a lodge. In most cases, the lodge is ingeniously built of branches and mud. Some beavers, especially adult male Beavers, tunnel into the banks of rivers, lakes or ponds for their den sites. In areas where trees do not commonly grow or currents are swift, females may also occupy bank dens.

Although the Beaver is not a fast mover, it more than compensates with its immense strength. It is not unusual for this solidly built rodent to handle and drag—with its jaws—a 9-kg piece of wood. The Beaver's flat, scaly tail, for which it is so well known, increases an animal's stability when it is cutting a tree, and it is slapped on the water or ground to communicate alarm.

The Beaver is well adapted to its aquatic lifestyle. It has valves that allow it to close its ears and nostrils when it is submerged, and clear membranes slide over the eyes. Because the lips form a seal behind the incisors, the Beaver can chew while it is submerged without having water or mud enter the mouth. In addition to its waterproof fur, the Beaver has a thin layer of fat to protect it from cold water, and the oily secretion it continually grooms into its coat keeps its skin dry.

RANGE: Beavers can be found wherever there is water, from the northern limit of deciduous trees south to northern Mexico. They are absent only from the Great Basin, deserts of the Southwest and extensive grassland areas devoid of trees.

Total Length: 90–120 cm
Tail Length: 28–53 cm
Weight: 16–30 kg

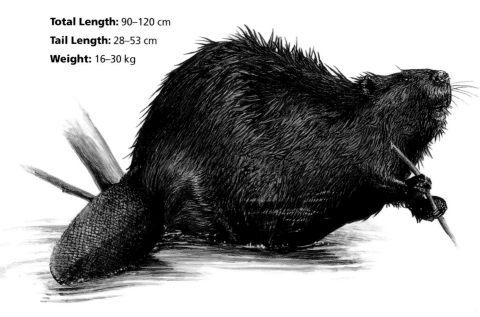

The Beaver is an impressive and industrious animal that shapes the physical settings of many wilderness areas. Although most tree cutting and dam building occur at dusk or at night, you may see the Beaver during the day—sometimes working, but usually sunning itself.

DESCRIPTION: The chunky, dark brown Beaver is the second-largest rodent in the world, taking a backseat only to the South American Capybara. It has a broad, flat, scaly tail, short legs, a short neck and a broad head, with short ears and massive, protruding, orange-faced incisors. The underparts are paler than the back and lack the reddish-brown hue. The nail on the next-to-outside toe of each webbed hindfoot is split horizontally, allowing it to be used as a comb in grooming the fur. The forefeet are not webbed.

HABITAT: Beavers occupy freshwater environments wherever there is suitable woody vegetation. They are some-times even found feeding on dwarf willows above treeline.

FOOD: Bark and cambium, particularly that of aspen, willow, alder and birch, is favoured, but aquatic vegetation is eaten in summer. Beavers sometimes come ashore to eat grains or grasses.

DEN: Beaver lodges are cone-shaped piles of mud and sticks. Beavers construct a great mound of material first, and then chew an underwater access tunnel into the centre and hollow out a den. The lodge is typically located away from shore in still water; in flowing

DID YOU KNOW?

Beavers are not bothered by lice or ticks, but there is a tiny, flat beetle that lives only in a Beaver's fur and nowhere else. This beetle feeds on Beaver dandruff, and its meanderings probably tickle sometimes, because Beavers often scratch themselves when they are out of water.

foreprint

walking trail

water it is generally on a bank. Access to the lodge is from about 1 m below the water's surface. A low shelf near the two or three plunge holes in the den allows much of the water to drain from the Beavers before they enter the den chamber. Beavers often pile more sticks and mud on the outside of the lodge each year, and shreds of bark accumulate on the den floor. Some "bank beavers" do not live in a lodge, but dig bank burrows at the water's edge. These burrows—the entrances to which are below water—may be as long as 50 m, but most are much shorter.

YOUNG: Most mating takes place in January or February, but occasionally as much as two months later. After a gestation of four months, a litter of usually four kits is born. A second litter may be born in some years. At birth, the 340–650-g kits are fully furred, their incisors are erupted, and their eyes are nearly open. The kits begin to gnaw before they are one month old, and weaning takes place at two to three months. Beavers become sexually mature when they are about two years old, at which time they often disperse from the colony.

Muskrat

SIMILAR SPECIES: The **Muskrat** (p. 150) is much smaller, and its long tail is laterally compressed rather than paddle-shaped. The **Northern River Otter** (p. 84) has a long, round, tapered, fur-covered tail, a streamlined body and a small head.

Eastern Chipmunk
Tamias striatus

These common chipmunks of southern Ontario are quite similar to their western counterparts. At one time, Eastern Chipmunks were the sole members of the genus *Tamias*, and the western species were classed as *Eutamius* species. Now all the chipmunks are grouped together in the genus *Tamias*. Like all chipmunks, Eastern Chipmunks have striped backs with five dark and four pale stripes. The outer two pale stripes are cream, while the two that border the dark dorsal stripe are greyish and often indistinct.

An Eastern Chipmunk nests in underground burrows and uses a tricky technique when creating its home. To excavate the main passageway and sleeping chamber of an underground nest, the chipmunk starts a "work hole." All the dirt excavated is strewn on the ground outside this work hole. When the burrow is finished, the work hole is closed and a main entrance is opened elsewhere. By excavating in this fashion, the main entrance has no mound of dirt to advertise its whereabouts. The tiny entranceway is known only to the chipmunk that built it.

With an insatiable drive, Eastern Chipmunks rapidly collect food for winter storage. For berries, they run along the twigs and quickly snip the stems with their sharp teeth. So speedy is their work that the ground beneath the berry bush seems bombarded with a storm of red pellets. Once enough berries are cut, the chipmunk scampers to the ground and carries the berries to a nearby cache. Most seeds, berries and nuts are collected in this manner. Throughout winter the chipmunk remains in its burrow. During wakeful periods of its hibernation, it eats the collected food items.

DESCRIPTION: This robust chipmunk differs from western chipmunks in its stripe pattern. The dorsal strip is quite narrow, and it runs from the back of the head to the rump. On either side of the dorsal stripe is a pale stripe that is either greyish or the same colour as the body. After this pale stripe there is a very dark stripe on each side, followed by a creamy yellow stripe and then another very dark stripe. The outermost dark stripes are much shorter than the others. The body coloration varies regionally, ranging from light oaky brown to dark walnut. Its undersides are quite pale, nearly white. There are two pale facial stripes: one above the eye and one below the eye, but overall the facial stripes are indistinct. The tail is relatively short; it is brown on top and

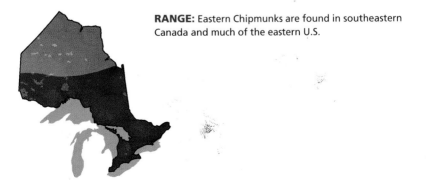

RANGE: Eastern Chipmunks are found in southeastern Canada and much of the eastern U.S.

Total Length: 23–30 cm
Tail Length: 7.2–10 cm
Weight: 66–139 g

edged with black. The ears are prominent and rounded.

HABITAT: Eastern Chipmunks are both "urban" and wilderness inhabitants. They live as comfortably in backyards and city parks as they do in open deciduous woodlands, forest edges, brushy areas and rocky outcroppings.

DEN: Primarily ground dwellers, Eastern Chipmunks excavate simple to complex burrows. Simple burrows are one main passageway with one storage chamber and one den chamber. Usually there is a secret hidden entrance. More complex burrows may have multiple passageways and entrances, several storage chambers and various smaller galleries or pockets in which to store debris or food. Occasionally, a female with her young may be found in a simple nest inside a hollow tree.

FOOD: Devoted gathers, these chipmunks continuously run between food sources and their storage chambers to store as much food as possible. Favourite items include berries, nuts, seeds and mushrooms. Perishable foods, such as slugs, insects and snails, are usually eaten right away. Like other chipmunks, Eastern Chipmunks will feed on carrion when the opportunity arises.

YOUNG: They are born either in a special nest chamber of the burrow or in a safe nest inside a hollow tree. Mating occurs in early spring, and the female gives birth to three to five young sometime in May. The young are altricial and require great care for several weeks. When they are one month old, they resemble small adults.

SIMILAR SPECIES: The smaller **Least Chipmunk** (p. 160) has more distinct stripes over the face, including a dark stripe through the eye, and its dorsal stripes continue to the base of the tail. No other chipmunks occur in Ontario.

DID YOU KNOW?

Although this chipmunk is essentially a ground- and shrub-dwelling species, it may also climb high into large oak trees in the autumn to get at the ripe acorns.

Least Chipmunk
Tamias minimus

The sound of scurrying among fallen leaves, a flash of movement and sharp, high-pitched "chips" are often enough to direct your attention to the fidgety behaviour of a Least Chipmunk. Using fallen logs as runways and the leaf litter as its pantry, this busy animal inhabits wooded areas throughout central Ontario. This chipmunk is less frequently seen in Ontario than the Eastern Chipmunk (p. 158).

The word "chipmunk" is thought to be derived from the Algonquian word for "head first," which is the manner in which a chipmunk descends a tree. Contrary to Disney-inspired myths, however, chipmunks spend very little time in high trees. They prefer the ground, where they bury food and dig golf ball–sized entrance holes to their networks of tunnels. Chipmunk burrows are known for their well-hidden entrances, which never have piles of dirt to give away their locations.

In certain heavily visited parks and golf courses, Least Chipmunks that have grown accustomed to human handouts can be very easy to approach. These exchanges contrast dramatically with the typically brief sightings of wild chipmunks, which scamper away at the first sight of humans. In the wild, chipmunks rely on their nervous instincts to survive in the predator-filled world.

DESCRIPTION: This tiny chipmunk has three dark and two light stripes on its face and five dark and four light stripes on its body. One of the dark stripes on its face is through the eye. The central dark stripe runs from the head to the base of the tail, but the other dark stripes end at the hips. The overall colour is greyer and paler than other chipmunks, and the underside of the tail is yellower. The coat of this chipmunk changes seasonally; in summer its coat is new and bright, and in winter its coat is duller, as if the chipmunk rolled in the dust to mute its colours. The tail is quite long—more than 40 percent of the total length.

HABITAT: The Least Chipmunk inhabits a wide variety of areas, including open coniferous forests, rocky outcroppings and pastures with small shrubs. It may be seen on farms well away from its normal habitat, attracted by livestock feed.

FOOD: This chipmunk loves to dine on ripe berries, such as chokecherries, pincherries, strawberries, raspberries or blueberries. Other staples in the diet include fungi, nuts, seeds, grasses and

RANGE: The extensive range of this species spreads from the central Yukon southeast to western Québec and south to northern California, New Mexico and North Dakota.

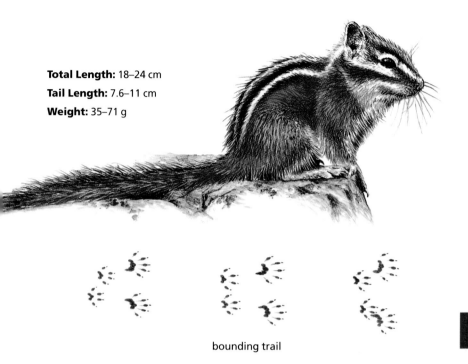

Total Length: 18–24 cm
Tail Length: 7.6–11 cm
Weight: 35–71 g

bounding trail

even insects and some other animals. It may be an important predator on eggs and nestling birds during the nesting season. A chipmunk is often attracted to domestic animal feed, and sometimes one will be filling its cheek pouches from a pile of oats shared by a horse.

DEN: The majority of Least Chipmunks den in underground burrows, which have concealed entrances, but some individuals live in tree cavities or even make spherical leaf and twig nests among the branches in the manner of tree squirrels.

YOUNG: Breeding occurs about two weeks after the chipmunks emerge from hibernation in spring. After about a one-month gestation, a litter of two to seven (usually four to six) helpless young is born in a grass-lined nest chamber. The young develop rapidly, and the mother may later transfer them to a tree cavity or tree nest.

SIMILAR SPECIES: The only other chipmunk in Ontario is the larger **Eastern Chipmunk** (p. 158), which has a shorter tail, lacks the stripe through the eye, and has shorter dorsal stripes.

DID YOU KNOW?

During summer, a chipmunk's body temperature is 35°–42° C. During winter, when it is hibernating in its burrow, its body temperature drops to 5°–7° C.

Woodchuck
Marmota monax

For much of the year, Woodchucks are tucked quietly away more than 2 m underground, relying on a lethargic metabolism during hibernation to keep them alive. They lie motionless, breathing an average of once every six minutes and maintaining life's requirements with a metabolic pilot light fed by a trickle of fatty reserves. Once late April returns (never as early as February's Groundhog Day), Woodchucks awake from their catatonic slumbers to breed and to forage on the palatable new green shoots emerging with the warmer weather.

Woodchucks range across all of Ontario, but they are much less common in northern areas. They find shelter for their burrows in rock piles, under old buildings and along riversides. In general, they are more solitary in nature than other kinds of marmots, and they are rarely seen far from their protective burrows, valuing security over the temptations of foraging. For feeding, Woodchucks tend to favour the early evening or shortly after dawn, but they can be seen at any time of the day. In most cases, they are wary and usually outrun intruders in an all-out sprint back to the burrow. A shrill whistle of alarm typically accompanies a Woodchuck's disappearance into its burrow.

Historically, the Woodchuck lived in forested areas. This mammal can still be found in open woodlands, but it now lives in great numbers on cultivated land—the Woodchuck is among the few mammals to have prospered from human activity. Unabashed about pilfering, Woodchucks that live near humans often graze in sweet alfalfa crops to help fatten their waistlines. The luckiest Woodchucks find their way into people's backyards, where they stuff themselves on tasty apples, carrots, strawberries and other garden delights.

ALSO CALLED: Groundhog, Marmot.

DESCRIPTION: This short-legged, stout-bodied, ground-dwelling marmot is brownish, with an overall grizzled appearance. It has a prominent, slightly flattened, bushy tail and small ears. Its feet and tail are dark or even black. Some individuals have whitish or tawny patches around the mouth.

HABITAT: Woodchucks favour pastures, meadows and open woodlands.

FOOD: In wild areas, this ground-dweller follows the standard marmot diet of grass, leaves, seeds and berries, which it supplements with bark and

RANGE: The Woodchuck occurs from central Alaska east to Labrador and south to northern Idaho in the West and eastern Kansas, northern Alabama and Virginia in the East.

Total Length: 46–66 cm
Tail Length: 11–16 cm
Weight: 1.8–5.4 kg

sometimes a bit of carrion. The Wood-chuck loves garden vegetables, and if it makes its way into urban areas, it may dine happily on corn, peas, apples, lettuce and melons.

DEN: The Woodchuck's powerful digging claws are used to excavate burrows in areas of good drainage. The main burrow is 3–15 m long, and it ends in a comfortable, grass-lined nest chamber. Plunge holes, without dirt piles, often lead directly to the chamber. A separate, smaller chamber is used for wastes.

YOUNG: Mating occurs in spring, within a week after the female emerges from hibernation. Some evidence suggests that a male may enter a female's den and mate with her before she arouses from hibernation. After a gestation of about one month, one to eight (usually three to five) young are born. The helpless newborns weigh only about 260 g. In four weeks their eyes

open, and they look like proper Wood-chucks after five weeks. The young are weaned at about 1½ months. Their growth accelerates once they begin eating plants, and they continue growing throughout summer to put on enough fat for winter hibernation and early spring activity.

SIMILAR SPECIES: The similar-sized **Muskrat** (p. 150) has a naked tail and is almost always associated with water; the **Franklin's Ground Squirrel** (p. 164) is smaller.

DID YOU KNOW?

Woodchucks are superb diggers that are responsible for turning over large amounts of earth each year. As they burrow, they periodically turn themselves around and bulldoze loose dirt out of the tunnel with their suitably stubby heads.

Franklin's Ground Squirrel
Spermophilus franklinii

Although these ground squirrels are known to live in a few places in Ontario, they are probably the most difficult squirrel in the province to observe. They live in edge communities where long grass or brushy areas meet woodlands, and they spend much of their time underground—adults that have emerged from hibernation in mid-April are typically back underground in August or early September, which amounts to an astonishing 7 to 8½ months of hibernation. Only the young of the year are still above ground in September; they need more time than adults do to attain their minimum hibernation weight. If you find this animal's burrows and wait patiently nearby, you may be rewarded with a good sighting as the ground squirrel runs off to a nearby food source.

With its large, bushy, grey tail and its affinity for shrubs and even open forests, this ground squirrel superficially resembles a tree squirrel. In fact, the resemblance is more than superficial; the Franklin's Ground Squirrel is often found within sprinting distance of a tree, and it is an active climber. Deciduous forests tend to be preferred, but in certain parts of central Canada this squirrel is found where conifers prevail. Where it is found in Ontario, this ground squirrel could easily be mistaken for the Eastern Grey Squirrel (p. 168).

The name of this ground squirrel honours the famed English Arctic explorer Sir John Franklin, who led three expeditions attempting to locate the Northwest Passage. (He finally lost his life, and the lives of his crew, in a failed attempt in 1846.) Interestingly, the first scientific collection of the Franklin's Ground Squirrel was by Sir John Richardson, a medical doctor who also served aboard Franklin's expeditions as a naturalist, and for whom another squirrel, the Richardson's Ground Squirrel (*S. richardsonii*), is named.

DESCRIPTION: This ground squirrel's most noticeable feature is its tail, which makes up about one-third of the animal's total length and is almost as bushy as a tree squirrel's. The tawny to olive tail is sprinkled with black and white hairs and has a white border. The overall body colour is grey, darker on the top of the head and lighter around the snout and on the sides of the face. A whitish ring surrounds the eye. The back is brownish grey, with indistinct light dapples and brownish transverse barring that becomes more pronounced on the rump. The undersides are grey or buffy coloured, and the feet are grizzled grey.

RANGE: Associated with the Prairies and Great Plains, the range of the Franklin's Ground Squirrel extends from central Alberta and central Saskatchewan southeast to Kansas and central Illinois.

Total Length: 33–43 cm
Tail Length: 12–16 cm
Weight: 360–700 g

HABITAT: Typically, these ground squirrels live in tall- and mid-grass prairies and brushy regions that border open woodlands or in grassy forest meadows, aspen bluffs or even the edges of dense coniferous forests.

FOOD: Grasses, green vegetation, berries and seeds make up about two-thirds of the diet; the remaining third is animal matter. These ground squirrels are effective predators that may take mice, young birds, eggs, frogs, toads, other ground squirrels and even small rabbits and ducks. They also feed on carrion.

DEN: This species is mainly solitary, but it may gather in small, loose colonies in good habitat with abundant food sources. The burrows are usually well concealed and may descend 1–2 m underground. The details of the burrow system are not well known, but presumably the design is similar to that of other ground squirrels, with multiple entrances and one main nest chamber.

The female lines her nest with grass. The burrow entrance usually lacks a spoil pile of dirt.

YOUNG: Mating occurs in the first week of May. After a gestation of about 4 weeks, a litter of 2 to 13 (usually 7 to 9) young is born. The newborns resemble pink gummy bears and are completely reliant on their mother. Their eyes open at 20 to 27 days, and by day 29 or 30 they are weaned and foraging by themselves. They reach adult size by mid-September.

SIMILAR SPECIES: The **Woodchuck** (p. 162) is larger.

DID YOU KNOW?

During the courtship season, rival males fight violently. The combatants bite one another, particularly on the genitalia, while the females await the outcome. The victorious male generally pursues the females through dense brush.

Red Squirrel
Tamiasciurus hudsonicus

Few squirrels have earned such a reputation for playfulness and agility as the Red Squirrel. This squirrel is a well-known backyard and ravine inhabitant that often has a saucy regard for its human neighbours. Like a one-man band, the Red Squirrel firmly scolds all intruders with shrill chatters, clucks and sputters, falsettos, tail flicking and feet stamping. Even when it is undisturbed, this loquacious squirrel often chirps as it goes about its daily routine. While the Red Squirrel is common in northern Ontario, it is rare in southwestern Ontario.

For this industrious squirrel, the daytime hours are devoted almost entirely to food gathering and storage. It urgently collects conifer cones, mushrooms, fruits and seeds in preparation for the winter months. The Red Squirrel remains active throughout winter, except in severely cold weather. At temperatures below −25° C it stays warm, but awake, in its nest.

Because the Red Squirrel does not hibernate, it needs to store massive amounts of food in winter caches. These food caches, which in extreme cases can reach the size of a garage, are the secret to the Red Squirrel's winter success. Most food caches are smaller than a cubic metre. Much of the Red Squirrel's efforts throughout the year are concentrated on filling these larders, and biologists speculate that its characteristically antagonistic disposition is a result of having to continually protect its food stores.

By the end of winter, Red Squirrels are ready to mate. Their courtship involves daredevil leaps through the trees and high-speed chases over the forest floor. Later, the youngsters are playful and frequently challenge nuts or mushrooms to a bout of aggressive mortal combat.

DESCRIPTION: The shiny, clove brown summer coat sometimes has a central reddish wash along the back. A black longitudinal line on each side separates the dorsal colour from the greyish to white underparts. There is a white eye ring. The backs of the ears and the legs are rufous to yellowish. The longest tail hairs have a black subterminal band and a buffy tip, which gives the tail a light fringe. The longer, softer winter fur tends to be bright to dusky rufous on the upperparts, with fewer buffy areas, and the head and belly tend to be greyer. The whiskers are black.

HABITAT: Boreal coniferous forests and mixed forests make up the major habitat, but towns with trees more than 40

RANGE: The Red Squirrel occupies coniferous forests across most of Alaska and Canada. In the West, it extends south through the Rocky Mountains to southern New Mexico. In the East, it occurs south to Iowa and Virginia and through the Alleghenies.

Total Length: 27–36 cm
Tail Length: 9–16 cm
Weight: 140–250 g

years old also support populations of Red Squirrels.

FOOD: Most of the diet consists of seeds extracted from conifer cones. A midden is formed where discarded cone scales and cores pile up below a favoured feeding perch. Flowers, berries, mushrooms, eggs, birds, mice, insects and even baby Snowshoe Hares or chipmunks may be eaten.

DEN: Tree cavities, witch's broom (created in conifers in response to mistletoe or fungal infections), logs and burrows may serve as den sites. The burrows or entrances are about 15 cm in diameter, with an expanded cavity housing a nest ball that is 40 cm across.

YOUNG: Northern populations bear just one litter a year. Peak breeding, in April and May, is associated with frenetic chases and multiple copulations lasting up to seven minutes each. After a 35- to 38-day gestation, a litter of two to seven (usually four or five) pink, blind, helpless young is born. The eyes open at four to five weeks, and the young are weaned when they are seven to eight weeks old. Red Squirrels are sexually mature by the following spring.

SIMILAR SPECIES: The **flying squirrels** (pp. 171–72) are a similar size, but they are a sooty pewter colour and nocturnal. The **Eastern Grey Squirrel** (p. 168) and the **Eastern Fox Squirrel** (p. 170) are both much larger.

DID YOU KNOW?

In a race against time, a Red Squirrel works to store spruce cones—as many as 14,000—in damp caches that prevent the cones from opening. If the cones open naturally on the tree, the valuable, fat-rich seeds are lost to the wind.

Eastern Grey Squirrel
Sciurus carolinensis

The Eastern Grey Squirrel is the most frequently encountered large squirrel in eastern North America, and although its range only includes southern Ontario, it can be locally abundant in both cities and natural areas. This squirrel is active throughout the entire year, sometimes even digging through snow to retrieve its buried nuts.

Many stories are told of the great migrations of grey squirrels, but squirrels do not actually migrate. In the autumn, young animals disperse, and in times of food shortage, adults may also disperse to find better homes. Technically, a migration is a movement of animals to and from specific regions in response to changing seasons (reproductive cycles may or may not be involved), whereas these squirrels "reshuffle" in response to food and population stresses. At times when large nut crops and high reproduction rates among the squirrels is followed by a year of little food, hundreds or even thousands of squirrels may move to find new food sources. When these squirrels travel, they can cover large tracks of forests without ever touching the ground.

The mainstay diet for these squirrels is nuts and seeds. Grey squirrels have dozens of nut caches, buried just under the surface of the soil. While the caches of most other squirrels germinate if left for too long, Eastern Grey Squirrels determinedly nip off the germinating end of the nuts before burying them. These squirrels routinely travel throughout their home range, keeping apprised of the fresh food sources available. Their caches of nuts are security for winter and stormy days; fresh corn crops, flowers, fruits, and mushrooms are relished in the present.

The Eastern Grey Squirrel is the classic "park and garden squirrel" of much of the Western world. This animal entertains millions of urban residents who restrict their wild adventures to places in cities. As a wildlife ambassador, this squirrel treats the urban world to a tempered, but genuine natural experience. This squirrel has even been introduced in Great Britain, where it inhabits city parks and natural areas and is known as the American Grey Squirrel. Introductions of animals into a new area is never without complications; where this squirrel is introduced it may have a negative effect on native songbirds by feeding on eggs and hatchlings in spring.

DESCRIPTION: The Eastern Grey Squirrel has two distinct colour forms. Some adults are dusty grey with pale undersides and a silvery, flattened tail. The

RANGE: This squirrel's native range encompasses all of the eastern U.S. and parts of Canada to southern Manitoba and Ontario in the north and eastern Texas in the south. Introduced populations have been established in Calgary, Vancouver, Victoria, Seattle and other western cities.

Total Length: 43–50 cm
Tail Length: 21–24 cm
Weight: 400–710 g

long tail hairs are white-tipped. Sometimes these squirrels have cinnamon highlights on their head, back and tail. The other form is solid black. In some localities, the entire squirrel population may be black. Albinos also occur, but they are very rare.

HABITAT: These squirrels prefer mature deciduous or mixed forests, with lots of nut-bearing trees. Older forests with larger trees support larger populations of squirrels—larger trees provide more food and numerous suitable nesting sites.

FOOD: These nut-lovers feed mainly on the seeds of oak, maple, ash and elm. In spring and summer, they also eat buds, flowers, leaves and occasionally animal matter, such as eggs or nestling birds.

DEN: Eastern Grey Squirrels den in trees year-round. They build nests lined with dry vegetation in natural tree cavities or in refurbished woodpecker holes. Where cavities are not available, they build dreys, which are spherical leaf and twig nests in tree branches. They have been known to make ground nests in cold regions, but this behaviour is not common.

YOUNG: Breeding occurs from December to February and rarely in July or August. Most females have only one litter a year. Gestation is 40 to 45 days, after which a litter of 1 to 8 (usually 2 to 4) helpless young is born. The eyes open at 32 to 40 days, and weaning occurs about three weeks later.

SIMILAR SPECIES: The slightly larger **Eastern Fox Squirrel** (p. 170) usually has more red in its fur (especially on the belly) and yellow-tipped hairs on the tail. It only occurs on Pelee Island. The **Red Squirrel** (p. 166) is smaller and reddish brown, and **flying squirrels** (pp. 171–73) are smaller, nocturnal and sooty grey.

DID YOU KNOW?

Eastern Grey Squirrels (as well as several other squirrel species) "live and learn"—they become wiser the older they get. Although many die before they are a year old, some may live quite a long time; a few individuals can make it to 10 years old or more.

Eastern Fox Squirrel
Sciurus niger

Total Length: 45–70 cm

Tail Length: 20–33 cm

Weight: 500–1050 g

This large and colourful tree squirrel is extremely rare in Ontario, and the only stable population lives on Pelee Island in Lake Erie. The Eastern Fox Squirrel usually leads a solitary life. Its most gregarious behaviour occurs in winter when several adults, whose ranges overlap and who are often related, share food caches and tree cavities. It remains active in winter, and the individuals sharing a tree cavity come and go regardless of the cold. The sharing that occurs between Eastern Fox Squirrels is not as sociable as the behaviour seen in other squirrels groups. The fox squirrels do not groom each other, "kiss" or nuzzle to maintain friendly ties.

DESCRIPTION: Although the Eastern Fox Squirrel has three different colour phases, only one phase is found here in Ontario: rusty orange or greyish above and cinnamon coloured on the undersides. The long tail hairs are reddish, as are the ears.

RANGE: This squirrel can be found in all of the eastern U.S. In Ontario, the only population is on Pelee Island.

HABITAT: These squirrels prefer mature mixed or deciduous forests, particularly oak/hickory forests. Older forests with larger trees support larger populations of squirrels—larger trees provide more food and numerous suitable nesting sites.

FOOD: Fox squirrels avidly feed on seeds, fruits, fungi, green pinecones and corn. Any non-perishables, like seeds and nuts, they bury in caches just under the ground surface. Like other squirrels, they may consume animal matter.

DEN: Fox squirrels den in trees year-round. They either build dreys (spherical leaf and twig nests), or they use natural tree cavities or woodpecker holes.

YOUNG: Breeding for this squirrel occurs in mid-winter, and a litter of two to four young is born in February or March. In ideal circumstances, these squirrels may have a second litter in August or September. The eyes of the young open at 32 to 40 days, and weaning occurs about three weeks later.

SIMILAR SPECIES: The slightly smaller **Eastern Grey Squirrel** (p. 168) is not as red and has white-tipped hairs on the tail. The **Red Squirrel** (p. 166) is smaller and reddish brown. **Flying squirrels** (pp. 171–73) are much smaller, nocturnal and sooty grey.

Southern Flying Squirrel
Glaucomys volans

Total Length: 20–25.5 cm

Tail Length: 8–12 cm

Weight: 45–100 g

FOOD: Like other squirrels, these mammals collect and eat nuts, seeds, fruits, mushrooms and insects. Their taste for meat is perhaps stronger than other squirrels because they regularly eat small vertebrates and carrion.

Small and steel grey, the Southern Flying Squirrel is a common resident of eastern forests. It is active after sunset and just before morning, so very few people ever see it.

Although these squirrels are primarily non-hibernating, severe winter conditions may induce a metabolic torpor in some of them. This torpid state outwardly resembles hibernation, but it is short lived and not as deep. After the cold snap has passed, a torpid squirrel rouses and resumes activity. Huddling is a more common response to cold weather—in winter, 5 to 50 flying squirrels are often found together in a tree cavity, relying on their body temperatures to warm the den space.

DESCRIPTION: This is a small tree squirrel with a fine, cool grey coat that usually has cinnamon highlights. The undersides are nearly or completely white, right to the base of the hairs. The glide membranes are black-edged and lay loose between the fore and hind legs when not in use. Its tail is broadly plumed and flattened.

HABITAT: These gliders are found in deciduous forests, especially beech-maple and oak-hickory. Here in Ontario, they are mainly associated with the Carolinian Forest.

DEN: Flying squirrels refurbish old corvid nests or most often occupy tree cavities. Sometimes flying squirrels make their own spherical dreys (nests) that are often occupied by several individuals for several generations.

YOUNG: Mating occurs in early spring, and females bear a litter of two to seven young after 41 days of gestation. The young are altricial and are not weaned until about seven weeks old. The young stay with the mother for the summer unless another litter is born.

SIMILAR SPECIES: The **Northern Flying Squirrel** is larger and often has more brown in its coat, and its belly fur is white tipped, but greyish at the base. The **Red Squirrel** (p. 166) is reddish brown overall. **Grey** and **fox squirrels** (pp. 168–70) are much larger. All non-flying tree squirrels lack the glide membranes on their sides.

RANGE: The Southern Flying Squirrel is found in most of the eastern U.S. and into southern Ontario, Québec and Nova Scotia.

Northern Flying Squirrel
Glaucomys sabrinus

Like drifting leaves, Northern Flying Squirrels seem to float from tree to tree in forests throughout much of Ontario. These arboreal performers are one of two species of flying squirrels in North America that are capable of distance gliding.

Although it is not capable of true flapping flight—bats are the only mammals to have mastered it—a flying squirrel's aerial travels are no less impressive, with extreme glides of up to 100 m. Enabling this squirrel to "fly" are its glide membranes—cape-like, furred skin extending down the length of the body from the forelegs to the hindlegs.

Before a glide, a squirrel identifies a target and maneuvers into the launch position: a head-down, tail-up orientation in the tree. Then, using its strong hindlegs, the squirrel propels itself into the air with its legs extended. Once airborne, it resembles a flying paper towel that can make rapid side-to-side maneuvers and tight downward spirals. Such control is accomplished by making minor adjustments to the orientation of the wrists and forelegs. On the ground and in trees, a flying squirrel hops or leaps, but the skin folds prevent it from running. It is not able to swim either.

The call of the Northern Flying Squirrel is a loud *chuck chuck chuck,* which increases in pitch to a shrill falsetto when the animal is disturbed. Like other tree squirrels, the Northern Flying Squirrel does not hibernate. On severely cold days, however, groups of 5 to 10 individuals can be found huddled in a nest to keep warm.

DESCRIPTION: Flying squirrels have a unique web or fold of skin that extends laterally to the level of the ankles and wrists to become the abbreviated "wings" with which the squirrel glides. These squirrels have large, dark, shiny eyes. The back is light brown, with hints of grey from the lead-coloured hair bases. The feet are grey on top. The belly hairs are white tipped, but grey or lead grey at the base. The underparts are light grey to cinnamon precisely to the edge of the gliding membrane and edge of the tail. The tail is noticeably flattened top to bottom, which adds to the buoyancy of the "flight" and helps the tail function as the rudder and elevators do on a plane.

HABITAT: Coniferous and mixed forests are prime flying squirrel habitat.

FOOD: The bulk of the diet consists of lichens and fungi, but flying squirrels also eat buds, berries, some seeds, a

RANGE: This flying squirrel occurs in eastern Alaska and across most of Canada in appropriate habitats. Its range extends south through the western mountains to Utah and California, around the Great Lakes and through the Appalachians.

Total Length: 25–38 cm
Tail Length: 11–18 cm
Weight: 75–180 g

few arthropods, bird eggs and nestlings and the protein-rich, pollen-filled male cones of conifers. They cache cones and nuts.

DEN: Nests in tree cavities, which are most commonly used, are lined with lichen and grass. Leaf nests, called dreys, are located in a tree fork close to the trunk. Twigs and strips of bark are used on the outside, with progressively finer materials used inside until the centre consists of grasses and lichens. If the drey is for winter use, it is additionally insulated to a diameter of 40 cm.

YOUNG: Mating takes place between late March and the end of May. After a six-week gestation, typically two to four young are born. They weigh about 5 g at birth. The eyes open after about 52 days. Ten days later they first leave the nest, and they are weaned when they are about 65 days old. Young squirrels

first glide at three months; it takes them about a month to become skilled. Flying squirrels are not sexually mature until after their second winter.

SIMILAR SPECIES: The **Southern Flying Squirrel** (p. 171) is smaller and is often more pewter coloured, and its belly hairs are white to the base. The **Red Squirrel** (p. 166) is reddish brown overall. **Grey** and **fox squirrels** (pp. 168–70) are much larger. All non-flying tree squirrels lack the glide membranes on their sides.

DID YOU KNOW?

Northern Flying Squirrels are often just as common in an area as Red Squirrels, but they are nocturnal and therefore rarely seen. Flying squirrels routinely visit bird feeders at night; they value the seeds as much as sparrows and finches do.

HARES & RABBITS

These rodent-like mammals are often called "lagomorphs" after the scientific name of the order, *Lagomorpha*, which means "hare-shaped." Rabbits, hares and pikas share the rodents' trademark, chisel-like upper incisors, and taxonomists once grouped the two orders together. Unlike rodents, however, lagomorphs have a second pair of upper incisors. Casual observers never see these peg-like teeth, which lie immediately behind the first upper incisor pair.

Lagomorphs are strict vegetarians, but they have relatively inefficient, non-ruminant stomachs that have trouble digesting such a diet. To make the most of their meals, they defecate pellets of soft, green, partially digested material that they then reingest to obtain maximum nutrition. Bacteria that enter the food in the intestines contribute to better digestion and absorption of nutrients the second time around. The pellets excreted after the second digestive process are brown and fibrous.

Hare Family (Leporidae)

Rabbits and hares are characterized by their long, upright ears, long jumping hindlegs and short, cottony tails. These timid animals are primarily nocturnal, and many spend the day resting in shallow depressions, called "forms." Rabbits and hares both belong to the same family, but they have distinct differences between them. Luckily, differentiating between rabbits and hares is easy, with the right information. The native hares turn white in winter, are larger overall, and tend to have longer ears and hindlegs than rabbits. The chief difference between the two involves reproduction. Rabbits build a maternity nest for their young, and when the young are born they are altricial, meaning they are helpless, blind and naked at birth. Hares, on the other hand, do not make a maternity nest, and their young are precocial, meaning they are well developed and fully furred and have open eyes. Soon after birth, the leverets (young hares) begin to feed on vegetation, and the mothers nurse them only once per day.

Snowshoe Hare

White-tailed Jackrabbit
Lepus townsendii

Total Length: 53–64 cm
Tail Length: 7–11 cm
Weight: 3–5 kg

The White-tailed Jackrabbit—included in Ontario's fauna from just a few specimens and hopeful future establishment—is a lean sprinter and the largest native hare in Ontario (the European Hare is slightly larger, but it is not native here). Records of this hare in Ontario exist from northwestern regions, where suitable habitat could easily support a population. No recent records have occurred to indicate its presence, but biologists would not be surprised if this jackrabbit were to expand its range into the province. If it were to turn up again, it would be in the suitable habitat between Lake of the Woods and Rainy Lake. Keep your eyes open—a positive sighting would be an exciting discovery.

DESCRIPTION: In summer, the upperparts of this large hare are light greyish brown, and the belly is nearly white. By mid-November, the entire coat is white, except for the greyish forehead and black-tipped ears. It has a fairly long, white tail that sometimes has a greyish band on the upper surface.

HABITAT: This hare prefers open areas, and it may be expanding its range northward, perhaps in association with land clearing. It will enter open woodlands to seek shelter in winter, but it avoids forested areas.

FOOD: Grasses and forbs make up most of the diet.

DEN: There is no true den, but a shallow form beside a rock or beneath sagebrush serves as a daytime shelter. In winter, jackrabbits may dig depressions or short burrows as shelters in snowdrifts.

YOUNG: One to nine (usually three or four) young are born in a form after a 40-day gestation. The fully furred newborns have open eyes and soon disperse, meeting their mother to nurse only once or twice a day. By two weeks they eat some vegetation; at five to six weeks they are weaned, often just before the birth of the next litter.

SIMILAR SPECIES: In winter, the more widespread but smaller **Snowshoe Hare** (p. 178) has lead grey, not creamy white, hair bases. The **Eastern Cottontail** (p. 180) is much smaller and does not turn white in winter. The **European Hare** (p. 176) is slightly larger and also does not turn white in winter.

RANGE: This hare is currently found from eastern Washington east to southern Manitoba and south to central California and eastern Kansas.

European Hare
Lepus europaeus

Have you ever wonder why the Easter Bunny—clearly a mammal—lays eggs? This story is not just a popular gimmick that sells Easter paraphernalia in spring, it is actually rooted in Germanic legend, and the European Hare is the source of this legend. According to the myth, the Goddess of Spring is Eostre, and she is a powerful goddess who created the hare (*Lepus europaeus*) by transforming a bird. Ever since this unusual conception, all hares have laid eggs during the week of Easter in gratitude to Eostre and in celebration of their ancestry.

From Germany, the European Hare has been widely introduced. The first introduction in Canada was in Ontario in 1912. Originally brought in as private game animals, they soon escaped their holds and bred freely. By 1919, these hares had established themselves in wild populations throughout the farmlands of southern Ontario.

European Hares that were introduced to many other countries around the world have easily adapted to different habitats. The hares released in Australia, for example, quickly adapted to the arid conditions, and their population exploded to pest proportions. Australians have tried many expensive but ineffective ways to eradicate this introduced mammal that over-grazes and out-competes native marsupials.

This athletic hare is an agile jumper and swift runner. When chased, the European Hare runs wildly in circles, zigzags and backtracks to confuse and elude its pursuer. When this hare is exerting itself to its limit, it can reach speeds of up to 75 km/h. Its ability to jump is remarkable; it is able to clear obstacles up to 1.5 m high, and it can make bounds up to 3.7 m long. Few animals can keep up with the European Hare, although foxes and Coyotes (p. 110) are agile and fast enough to be its major predators.

DESCRIPTION: This large hare is tawny to grey brown in colour, sometimes with flecks of black, and the undersides are white. Unlike the Snowshoe Hare (p. 178), this hare does not turn white in winter. The winter coat is mainly grey with white below. Its fur has a unique kinky or rough appearance and is very thick. The short tail is black on top and white below. It has long ears that are the same colour as the back.

HABITAT: These adaptable hares live in a variety of habitats. They prefer open meadows, golf courses or agricultural land that are close to woodlands and hedgerows.

RANGE: The European Hare is found in southern Ontario and east to New York and New England.

Total Length: 60–75 cm
Tail Length: 7–10 cm
Weight: 2.7–5.4 kg

FOOD: Strictly vegetarian, hares feed on clover, grass, grain, fruits and vegetables, and green vegetation. They exhibit "coprophagy" as a means of maximizing nutrition. Hares feed very quickly in the open to limit their exposure to predators, and then return to a safe place to reingest pellets and digest the vegetation properly.

DEN: Like other hares, European Hares sleep at night in forms. A female with young makes several separate forms hidden from view. She distributes her litter amongst the forms—one young per form. She visits each form in rotation at night to feed her young.

YOUNG: Mating usually peaks in spring, but it can occur at any time of the year. After a gestation of 30 to 40 days, the female gives birth to 2 to 4 young. Females may have up to four litters per year. The young are born with fur and with their eyes open. Although the young are quite developed, they nurse for up to three weeks. The mother weans them, and they disperse by the fourth week.

SIMILAR SPECIES: The other common hare in Ontario, the **Snowshoe Hare** (p. 178), is smaller and turns almost completely white in winter. The **White-tailed Jackrabbit** (p. 175) may occur in extreme western Ontario, and it has larger ears and also turns white in winter. The **Eastern Cottontail** (p. 180) does not turn white in winter, is smaller and has much smaller ears.

DID YOU KNOW?

When a European Hare makes a form to rest in, it must have a good view and receive a gentle breeze. The breeze carries sound and odour, so the hare can detect if a predator approaches.

Snowshoe Hare
Lepus americanus

The Snowshoe Hare is able to withstand the most unforgiving aspects of northern wilderness because it possesses several fascinating adaptations for winter. As its name implies, for example, the Snowshoe Hare has very large hindfeet and can easily walk on areas of soft snow, whereas other animals sink into the powder. This ability is a tremendous advantage for an animal that is preyed upon by so many different species of carnivores. Unfortunately, it is of minimal help against the equally big-footed Canada Lynx (p. 52), a specialized hunter of the Snowshoe Hare.

It is well known that populations of lynx and hares fluctuate in close correlation with one another, but few people realize that other species are involved in the cycle. Recent studies have shown that as hares increase in number, they overgraze willow and alder in their habitat. These plants are their major source of food during winter months. In response to overgrazing, the willow and alder produce a distasteful and toxic substance in their shoots that is related to turpentine. This substance protects the plants, but it initiates starvation in the hares. As the hares decline, so do the lynx. Once the plants recover their growth after a season or two, their shoots become edible again and the hare population increases.

In response to shorter daylight at the onset of winter, Snowshoe Hares start moulting into their white winter camouflage whether snow falls or not. The hares have no control over the timing of this transformation. If the year's first snowfall is late, some individuals will lose their usual concealment and become visible from great distances—to naturalists and predators alike—as bright white balls in a brown world. The hares seem to be aware of this predicament, and they often seek out any small patch of snow on which to squat.

DESCRIPTION: The summer coat is rusty brown above, with the crown of the head darker and less reddish than the back. The nape of the neck is greyish brown, and the ear tips are black. The chin, belly and lower surface of the tail are white. Adults have white feet; juveniles have dark feet. In winter, the terminal portion of nearly all the body hair becomes white, but the hair bases and underfur are lead grey to brownish. The ear tips remain black.

HABITAT: Snowshoe Hares may be found in much of Ontario, primarily where there is forest or dense shrubs.

RANGE: The range of the Snowshoe Hare is associated with the boreal coniferous forests and mountain forests from northern Alaska and Labrador south to California and New Mexico.

Total Length: 38–53 cm
Tail Length: 4.8–5.4 cm
Weight: 1–1.5 kg

FOOD: In summer, a wide variety of grasses, forbs and brush may be consumed. In winter, mostly the buds, twigs and bark of willow and alder are eaten. These hares occasionally eat carrion.

DEN: Snowshoe Hares do not keep a customary den, but they sometimes enter hollow logs or the burrows of other animals; in areas of human activity they may run beneath buildings.

YOUNG: Breeding activity begins in March and continues through August. After a gestation of 35 to 37 days, 1 to 7 (usually 3 or 4) young are born under cover, but often not in an established form or nest. The female breeds again within hours of the birth, and she may have as many as three litters in a season. The young hares are precocial and can hop within a day; they feed on grassy vegetation within 10 days. In five months they are fully grown.

SIMILAR SPECIES: The **White-tailed Jackrabbit** (p. 175) has longer ears and a slightly longer tail, and its winter underfur is creamy white. The **Eastern Cottontail** (p. 180), which does not turn white in winter, is generally smaller and has an orangish or rusty nape of the neck, and its ears are generally uniform in colour or have white edges. The **European Hare** (p. 176) is larger and also does not turn white in winter.

DID YOU KNOW?

Between their highs and lows, Snowshoe Hare densities can vary by a factor of 100. During highs, there may be 12 to 15 hares per hectare; after a population crashes, hares may be relatively uncommon over huge geographical areas for years.

Eastern Cottontail
Sylvilagus floridanus

The Eastern Cottontail is native only to the eastern half of North America. These rabbits have been widely introduced throughout many of the western states, and it is now the most widespread cottontail in North America. Its success is partly because of its ability to adapt to a wide variety of habitats. The main requirement is tall grasses or shrubs that provide adequate cover for protection from predators. This cottontail has high fecundity, and although accurate population figures are not known for Ontario, its numbers are probably increasing.

During twilight, Eastern Cottontails emerge from their daytime hideouts to graze on succulent vegetation. If you are a patient observer, you may see them as they daintily nip at grasses, always just a short leap from dense bushes or a rocky shelter. Sometimes cottontails can be seen during the day, especially young ones or cottontails in cities.

The prime habitat for an Eastern Cottontail is neither fully wooded areas nor completely open flats—they require good protective cover to hide from predators, but in areas where foliage is too dense they are handicapped in their ability to detect an approaching predator. They are preyed upon heavily by Coyotes (p. 110), foxes, owls and hawks.

Eastern Cottontails spend most of their days sitting quietly in forms beneath impenetrable vegetation or under boards, rocks, abandoned machinery or buildings. These mid-sized herbivores have small home ranges about one to two hectares in size. Heavy rains greatly diminish cottontail activity, restricting them to their hideouts for the duration of the storm. Eastern Cottontails do not hibernate during winter, but they limit their movements to traditional trails that they can easily locate after a snowfall. In very cold winters, they have been known to take shelter in Woodchuck burrows.

DESCRIPTION: This rabbit is pale buffy grey above, with somewhat paler sides. Sometimes the back is grizzled with blackish hairs. The nape of the neck is orangish, and the legs are cinnamon in colour. The undersides are whitish. The tail is brown above and cottony white below, but the white only shows when the animal is running.

HABITAT: The only major habitat requirement is cover, whether it is brush, rocky outcroppings or buildings. This rabbit likes edge situations where trees meet meadows or where brushy areas meet agricultural land. It also

RANGE: This rabbit is found in the eastern U.S. and along the southern borders of several eastern Canadian provinces; introduced populations live in California, Oregon, Washington and B.C.

Total Length: 40–45 cm

Tail Length: 4–7 cm

Weight: 0.8–1.6 kg

inhabits brushy riparian areas, and it is occasionally seen on lawns, city parks or golf courses with nearby shrubby cover.

FOOD: This species favours clover, grasses and herbaceous vegetation in the summer and woody species in winter. It may also eat the bark of young trees.

DEN: Brush piles, holes and leaf litter are used for escape cover, but there is no true den other than for pregnant females. Before giving birth, the female digs a nest about 25 cm long and 15 cm wide and lines it with grasses and her own fur; she then covers the depression so it is nearly impossible to see. The nest is essentially invisible, and a casual observer would never suspect that the female was nursing, or that a nest of babies lay beneath her.

YOUNG: These rabbits are among the most fecund of all rabbits, and they may breed at anytime of the year in suitable climates. Following mating and a gestation of 18 to 30 days, 1 to 9 bunnies are delivered in the nest. Within hours of giving birth, the mother is in estrus again and can breed. At one month the young are independent, and as early as four months the young females can breed.

SIMILAR SPECIES: Both the **Snowshoe Hare** (p. 178) and the **White-tailed Jackrabbit** (p. 175) are much larger and become white in winter. The **European Hare** (p. 176) is larger and has longer ears.

DID YOU KNOW?

No other rabbit or hare is as widespread as the Eastern Cottontail. In fact, this species' range overlaps with seven other species of cottontails and six species of hares.

BATS

In an evolutionary sense, bats are a very successful group of mammals. Worldwide, nearly a quarter of all mammalian species are bats, and they are second only to rodents in both diversity of species and number of individuals. Unfortunately, across North America, populations of several bat species appear to be declining.

A bat's wing consists of a double layer of skin stretched across the modified bones of the fingers and down to the legs. A small bone, the calcar, juts backward from the foot to help support the tail membrane, which stretches between the tail and each leg. The calcar is said to be keeled when there is a small projection of skin from its side.

Bats generate lift by pushing their wings against the air's resistance, so they tend to have large wing surface areas for their body size. This method of flight is less efficient than the airfoil lift provided by bird or airplane wings, but it allows bats to fly slower and gives them more maneuverability. Slower flight is a real advantage when trying to catch insects.

Bats have good vision, but their nocturnal habits have led to an increased dependence on their sense of hearing—most people are acquainted with the ability of many bat species to navigate or capture prey in the dark using echolocation. The tragus—a slender lobe that projects from the inner base of many bats' ears—is thought to help in determining an echo's direction.

Bats, unlike most other types of small mammals, have small litters. The high energy requirements for flight limit most female bats to having only one offspring a year. Bat populations can reach high numbers, however, because bats live for many years.

Evening Bat Family (Vespertilionidae)

All eight species of bats that occur in Ontario belong to this family. True to their name, most members of this family are active in the evening and often again before dawn. A few species migrate to warmer regions for winter, but most hibernate in caves, abandoned mines or even buildings.

Key to the Bats of Ontario

1a. Dorsal surface of uropatagium densely furred 2

1b. Dorsal surface of uropatagium not densely furred 3

2a. Forearm less than 44 mm, fur strongly reddish ....*Lasiurus borealis* (p. 187)

2b. Forearm greater than 50 mm, fur grey with whitish tips ....*Lasiurus cinereus* (p. 188)

3a. Fur black with white tips ....*Lasionycteris noctivagans* (p. 190)

3b. Fur not distinctly black with white tips 4

4a. Forearm greater than 40 mm ....*Eptesicus fuscus* (p. 191)

4b. Forearm less than 40 mm 5

5a. Fur on venter same as on dorsal, dorsal fur tricoloured ....*Pipistrellus subflavus* (p. 192)

5b. Fur on venter not the same as fur on dorsal 6

6a. Forearm less than 34 mm, hindfoot less than 9 mm. ....*Myotis leibii* (p. 186)

6b. Forearm greater than 34 mm, hindfoot greater than 9 mm ..7

7a. Tragus long and pointed, splinter-like ....*Myotis septentrionalis* (p. 183)

7b. Tragus short and blunt ....*Myotis lucifugus* (p. 184)

Northern Bat
Myotis septentrionalis

Total Length: 83–100 mm
Tail Length: 29–45 mm
Forearm: 33–40 mm
Weight: 3.5–8.9 g

The Northern Bat tends to roost in natural cavities and under peeling or lifting bark on old or dead trees during the warmer months in Ontario. Some people are concerned that it is therefore vulnerable to forestry operations, which often select older trees for harvesting. The Northern Bat is considered a "gleaner," because when feeding it grabs insects off branches, leaves and other surfaces instead of catching them in flight.

ALSO CALLED: Northern Long-eared Bat.

DESCRIPTION: This mid- to dark brown bat has a wingspan of 23–25 cm. It has somewhat lighter underparts. A distinct dark "mask" is on the face. The tips of its hairs are lighter than the bases, giving a sheen to the fur. The tragus is long and splinter-shaped. The ears extend past the tip of the nose if pushed forward. The calcar is not keeled.

HABITAT: The Northern Bat occurs primarily in forested and sometimes brushy areas, and it prefers to be close to waterbodies.

FOOD: This bat feeds at dusk and again just before dawn. It catches small insects, especially flies and mosquitos.

DEN: In September and October, this mainly solitary bat seeks out caves and mines in which to hibernate. Despite being mainly solitary, the females form nursery colonies in spring. These colonies are usually located in tree cavities, under loose bark on trees, occasionally in old buildings, under bridges or in loose shingles on rooftops, and they may contain up to 30 females, but smaller groups are more common.

YOUNG: These bats mate in autumn, but fertilization is delayed until spring, so the single young is not born until June or early July, after a gestation of about 40 days. The young are able to fly in about four weeks.

SIMILAR SPECIES: All the mouse-eared bats (see *Myotis* spp, pp. 183–86) are essentially impossible to identify in flight. Even in hand, one needs a technical key (p. 182). The slightly smaller **Eastern Pipistrelle** (p. 192) has shorter, paler ears.

RANGE: Northern Bats are found from eastern B.C. east to Newfoundland and south to Nebraska, Arkansas, western Georgia and Virginia.

Little Brown Bat
Myotis lucifugus

On nearly every warm, calm summer night, the skies of Ontario are filled with marvellously complex screams and shrill chirps. Unfortunately for people interested in the world of bats, these magnificent vocalizations occur at frequencies higher than our ears can detect. The most common of these nighttime screamers, and quite likely the first bat most people will encounter, is the Little Brown Bat.

Once the cool days of late August and September arrive, Little Brown Bats begin to migrate to wintering areas. While it is not known where all of these bats spend winter, thousands of them travel to natural caves and abandoned mine shafts. Large wintering populations are known to occur in certain large caves, and cave adventurers are advised to take special care not to disturb these hibernating animals. All bats may rouse on occasion during hibernation, and some may even fly out on warm nights, but if a bat is roused too often by disturbance or intruders it risks speeding up its metabolism too much and running out of reserves before the arrival of insects in spring.

DESCRIPTION: As its name suggests, this bat is little and brown. Its coloration ranges from light to dark brown on the back, with somewhat paler undersides. The tips of the hairs are glossy, giving this bat a coppery appearance. The wing and tail membranes are mainly unfurred, though fur may appear around the edges. The calcar of this bat is long and unkeeled. The tragus, which is bent, is half the length of the ear. The wingspan is 22–27 cm.

HABITAT: Little Brown Bats are the most frequently encountered bats in much of North America. At home almost anywhere, you may find them in buildings, attics, roof crevices and loose bark on trees or under bridges. Wherever these bats are roosting, waterbodies are sure to be nearby. They need a place to drink and a large supply of insects for their nightly forays.

FOOD: Little Brown Bats feed exclusively on night-flying insects, especially new aquatic emergents. In the evening, these bats leave their day roosts and swoop down to the nearest water source to snatch a drink on the wing. Foraging for insects can last for up to five hours. Later, the bats take a rest in night roosts (a different place from their day roosts). Another short feeding period occurs just prior to dawn before the bat returns to its day roost.

RANGE: This widespread bat ranges from central Alaska to Newfoundland and south to northern Florida and central Mexico, except for much of the southern Great Plains.

Total Length: 70–100 mm
Tail Length: 25–54 mm
Forearm: 35–41 mm
Weight: 5.3–8.9 g

DEN: These bats may roost alone or in groups. Females in maternity colonies can number a few to 1000, and males roost singly or in small groups. A loose shingle, an open attic or a hollow tree are all suitable roosts for Little Brown Bats. In winter, these bats stay and hibernate in large numbers in caves and old mines.

YOUNG: Mating occurs either in late autumn or in the hibernation colonies. Fertilization of the egg is delayed until the female ovulates in spring, and by June, pregnant females form nursery colonies in a protected location. In late June or early July, one young is born to a female after about 50 to 60 days of gestation. The young are blind and hair-less, but their development is rapid and their eyes open in about three days. After one month, the young are on their own.

SIMILAR SPECIES: All the mouse-eared bats (*Myotis* spp, pp. 183–86) are essentially impossible to identify in flight. Even in hand, one needs a technical key (p. 182). The **Eastern Pipistrelle** (p. 192) has tricoloured hairs on its back.

DID YOU KNOW?

An individual Little Brown Bat can consume 900 insects an hour during its nighttime forays. A typical colony may eat 45 kg of insects a year.

Eastern Small-footed Bat
Myotis leibii

Total Length: 72–84 mm
Tail Length: 30–39 mm
Forearm: 30–34 mm
Weight: 3–8 g

branes and tail membrane are dark brown and the ears are black. Its wingspan is 21–25 cm, and some fur may be found on both the undersurface of the wing and the upper surface of the tail membrane. Across its face, from ear to ear, is a dark brown or black "mask." True to its name, this bat has noticeably small feet, 6–8 mm in length. The calcar is strongly keeled, and the tragus is long and pointed.

The Eastern Small-footed Bat is a very small and understudied member of this genus. This bat appears to be extremely rare, and throughout its range its distribution is thin and spotty. Wherever this bat is found, however, it is always near water such as streams and ponds. When this bat emerges at dusk to forage, it flies above the nearby water-bodies to capitalize on the abundance of night-flying insects over the water. This sensitive species may show decline when forests are removed from around their hibernation sites or foraging areas.

DESCRIPTION: The glossy fur of this attractive bat is yellowish brown to grey or even coppery brown above, and its undersides are buffy or whitish. The flight mem-

HABITAT: The Eastern Small-footed Bat prefers hilly regions with deciduous or evergreen forests, and it sometimes occurs in somewhat open farmland and grassy areas.

FOOD: Like most bats in Ontario, the Eastern Small-footed Bat eats primarily small flying insects.

DEN: In summer, this bat roosts under rock ledges, under bridges (in expansion joints) and occasionally in or under the eaves of buildings. In winter it hibernates either singly in cracks and crevices or in a group in caves and mines. Nursery colonies occur in bank crevices, under bridges or under the shingles of old buildings.

YOUNG: In small nursery colonies, one young per female is born from June or July.

SIMILAR SPECIES: All the mouse-eared bats (*Myotis* spp, pp. 183–86) are essentially impossible to identify in flight. Even in hand, one needs a technical key (p. 182). The similarly sized **Eastern Pipistrelle** (p. 192) lacks the black mask.

RANGE: The Eastern Small-footed Bat has a spotty distribution from southern Ontario and the northeastern states through the Appalachians to Kentucky and Missouri.

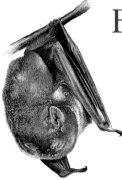

Eastern Red Bat
Lasiurus borealis

Total Length: 87–126 mm
Tail Length: 45–62 mm
Forearm: 37–42 mm
Weight: 7–15 g

This pretty bat is a close cousin of the more common Hoary Bat (p. 188). Considering that many other bats in Ontario have brownish fur, the Eastern Red Bat stands out and should be easy to recognize. Unfortunately, this bat begins foraging as late as two hours after sunset, at a time when it is too dark to recognize its reddish hue. Additionally, this bat is one of the solitary bats and keeps itself well hidden, so the chances of seeing one are slim. If you do find one, you may be able to identify whether it is male or female because males are much brighter red. On occasion, you might see one feeding on insects attracted to a streetlight, and the flush of light should allow you to see the colour.

DESCRIPTION: This medium to large bat has mainly yellowish-orange to red fur. The male is often brighter than the female. Some individuals may have a slightly frosted appearance owing to white-tipped hairs. The wingspan is 29–33 cm. The ears are small and rounded, and the tragus is small. The backsides of the ears and face are covered in orangish fur, and the upper surface of the tail membrane is furred. *Lasiurus* bats are unique in this region in having four mammae; all other bats here have two.

HABITAT: This bat lives in or near forests, both deciduous and coniferous and often in range of open, grassy areas.

FOOD: When it forages near farmlands, the Eastern Red Bat may feed heavily on agricultural pests. This bat primarily eats moths, plant hoppers, flies and beetles, and it may sometimes alight on vegetation to pick off insects. The peak feeding period is well after dusk.

DEN: In summer, these solitary bats roost in branches, which provides shade. The space beneath the roost must be free of obstacles to allow the bats to drop into flight. Beginning in autumn, they migrate south for winter.

YOUNG: Mating takes place in August and September, but ovulation and fertilization are delayed until spring. Gestation appears to be 80 to 90 days, and one to four young are born in June. They are thought to be able to fly when three or four weeks old and are weaned at five or six weeks. Their age at sexual maturity is not known.

SIMILAR SPECIES: The larger **Hoary Bat** (p. 188) has a frosty appearance over its back. The **Big Brown Bat** (p. 191) is dark brown, and other *Myotis* bats are smaller and also brown.

RANGE: The Eastern Red Bat ranges across much of southern Canada and most of the eastern U.S.

Hoary Bat
Lasiurus cinereus

The Hoary Bat is the largest bat in Ontario, with a wingspan of about 40 cm, but it still weighs less than the smallest chipmunk. It flies later into the night than any other bat in the region; once the last of the daylight has disappeared, the Hoary Bat courses low over wetlands, lakes and rivers in conifer country. It may not be as acrobatic in its foraging flights as the smaller *Myotis* bats, but no one who has ever witnessed a Hoary Bat in flight could fail to be impressed by its aerial accomplishments.

The large size of the Hoary Bat is often enough to identify it, but the light wrist spots, which are sometimes visible at twilight, confirm the identification. Many of the Hoary Bat's long hairs have brown bases and white tips, giving the animal a frosted appearance and its common name. Though attractive, this coloration makes the Hoary Bat very difficult to notice when it roosts in a tree—it looks very similar to dried leaves and lichens.

Hoary Bats, as well as other tree-dwelling bats, have been the focus of scientific study recently to determine the importance of old roost trees in their habitat. These bats have complex requirements: while old trees may well be important, water quality and the availability of hatching insects in wetlands may be equally significant.

The few records from the northern parts of Canada suggest that female Hoary Bats may migrate quite far north. The males, it is thought, migrate only as far as the northern U.S., where they likely court and mate with the females. While the males may remain at these sites for summer, some impregnated females appear to push farther north, where the young are born.

DESCRIPTION: The large Hoary Bat has greyish fur, and the white hair tips give it a heavily frosted appearance. Its throat and shoulders are buffy yellow or toffee coloured. Its wingspan is 38–41 cm. The ears are short, rounded and furred, but the edges of the ears are naked and black. The tragus is blunt and triangular. The upper surfaces of the feet and tail membrane are completely furred. The calcar is modestly keeled. Like the Eastern Red Bat, the Hoary Bat has four mammae.

HABITAT: The Hoary Bat is often found near open, grassy areas in coniferous and deciduous forests or over lakes. It is also common in cities, and it can be seen feeding at streetlights at night.

FOOD: The diet consists mainly of moths, plant hoppers, flies and beetles.

RANGE: From north-central Canada, the Hoary Bat ranges south through most of southern Canada and almost all of the lower U.S.

Total Length: 110–150 mm

Tail Length: 41–67 mm

Forearm: 45–57 mm

Weight: 19–35 g

When this bat forages over farmland, it consumes high numbers of agricultural pests. It sometimes alights on vegetation to pick off insects. Feeding activity does not peak until well after dusk.

DEN: This migratory bat usually returns to Ontario in May. During summer, it roosts alone in the shade of foliage, with an open space beneath the roost so that it can drop into flight. Beginning in August or September, it migrates south, sometimes in large flocks or "clouds."

YOUNG: Hoary Bats mate in autumn, but the young are not born until late May or June because fertilization is delayed until the female ovulates in spring. Gestation lasts about 90 days, and a female, which has four mammae, usually bears two young. She places the first young on her back while she delivers the next. Before they are able to fly, young bats roost in trees and nurse between their mother's nighttime foraging flights.

SIMILAR SPECIES: The **Eastern Red Bat** (p. 187) is smaller and has distinctly red fur. The **Silver-haired Bat** (p. 190) is black with silver-tipped hairs and is slightly smaller. The **Big Brown Bat** (p. 191) is almost as large, but it does not have a frosted appearance.

DID YOU KNOW?

The Hoary Bat is the most widespread species of bat in North America, and it is the only "terrestrial" mammal native to the Hawaiian Islands.

Silver-haired Bat
Lasionycteris noctivagans

Total Length: 90–110 mm
Tail Length: 35–51 mm
Forearm: 38–45 mm
Weight: 7–18 g

DESCRIPTION: The fur is nearly black, with long, white-tipped hairs on the back giving it a frosty appearance. The naked ears and tragus are short, rounded and black. The wingspan is 28–30 cm. A light covering of fur may be seen over the entire surface of the tail membrane.

The handsome Silver-haired Bat flies slowly and leisurely during twilight hours throughout central and southern Ontario. Twilight actually happens twice in a 24-hour period: once after sunset (vesperal twilight) and again before sunrise (auroral twilight).

The feeding forays of this bat also happen twice a day. They usually fly fairly low to the ground, and they don't seem to be disturbed by the presence of an inquisitive human. If you happen to see one, either at night or in very early morning, you may be able to watch it for some time as it dips and flops about the twilight sky catching insects. Listen closely as well, for you may hear the subtle *wicka-wicka-wicka* of its leathery wings.

HABITAT: Forests are the primary habitat, but this bat can easily adapt to parks, cities and farmlands.

FOOD: This bat feeds mainly on moths, and it forages over standing water or in open areas near water.

DEN: The summer roosts are usually in tree cavities, under loose bark or in old buildings. In winter these bats migrate out of the province. Females form nursery colonies in protected areas, such as tree cavities, narrow crevices or old buildings.

YOUNG: Breeding takes place in autumn, but fertilization is delayed until the female ovulates in spring. In early summer, after a gestation of about two months, one or two young are born to each female.

SIMILAR SPECIES: The Silver-haired Bat's white-tipped black hairs are unique among the bats of the region. The **Big Brown Bat** (p. 191) has mainly brown, glossy fur. The **Hoary Bat** (p. 188) does not have black fur.

RANGE: This bat is found along the southeastern coast of Alaska, across the southern half of Canada and south through most of the U.S.

Big Brown Bat
Eptesicus fuscus

Total Length: 90–140 mm

Tail Length: 20–60 mm

Forearm: 46–54 mm

Weight: 12–28 g

farmlands, it feeds heavily on agricultural pests. Foraging usually occurs at heights of no more than 9 m, and the two peak feeding periods are at dusk and just before dawn.

The Big Brown Bat is not overly abundant anywhere, but its habit of roosting and occasionally hibernating in houses and other human structures makes it a more commonly encountered bat. It is also the only bat that may be seen, rarely, on warm winter nights, because it occasionally takes such opportunities to change hibernating sites. The relative frequency of Big Brown Bat sightings doesn't save this species from the anonymity that plagues most bats, however, because the "big" in its name is relative—this sparrow-sized bat still looks awfully small against a dark night sky.

DESCRIPTION: This big bat is mainly brown, with lighter undersides, and its fur appears glossy or oily. On average, a female is larger than a male. The face, ears and flight membranes are black and mainly unfurred. The blunt tragus is about half as long as the ear. The calcar is usually keeled.

HABITAT: This large bat easily adapts to parks, cities, farmlands and buildings. In the wild, it typically inhabits forests.

FOOD: A fast flier, the Big Brown Bat feeds mainly on beetles and plant hoppers, but rarely moths or flies. Near

DEN: In summer, this bat usually roosts in tree cavities, under loose bark or in buildings. It spends winter hibernating in caves, mines or old buildings. Nursery colonies are found in protected areas, such as tree cavities, large crevices or old buildings.

YOUNG: These bats breed in autumn or during a wakeful period in winter, but fertilization is delayed until the female ovulates in spring. A female gives birth to one or two young in early summer after about a two-month gestation. As in most bats, the female has two mammae.

SIMILAR SPECIES: The Big Brown Bat is not easy to distinguish from the other large bats, but the **Hoary Bat** (p. 188) has frosted brown or grey fur, and the **Silver-haired Bat** (p. 190) has frosted black fur. The *Myotis* bats (pp. 183–86) are all smaller.

RANGE: This bat occurs from the southern tier of Canada through most of the U.S.

Eastern Pipistrelle
Pipistrellus subflavus

The Eastern Pipistrelle is not widespread in Ontario, but it may be locally abundant. In the southeastern parts of the province, it may be the first bat many people see because this bat begins foraging in the evening while there is still sunlight, and in the morning it feeds well after dawn. During most of the night, it rests, having ceased its foraging activity at about 9:30 p.m. This bat forages heavily throughout summer to develop a layer of fat. In Ontario, pipistrelles move into mines or caves where they hibernate. During their hibernation, they require the fat layer for sustenance.

Like other bats, Eastern Pipistrelles are very clean. After foraging each night, they spend as much as 30 minutes grooming their fur and cleaning out debris or bugs that have accumulated. They use their tongues wherever they can reach, and otherwise, like a cat, they moisten their hindfeet to clean the remaining areas. Special attention is given to cleaning the ears—bats are dependent on their hearing to "see" the world with echolocation, so a dirty ear would be intolerable. Most bats have good eyesight, too, but as nighttime fliers, hearing is more useful.

The flight of the Eastern Pipistrelle is weak and erratic. Its slow flight is advantageous to it when enthusiastic naturalists are attempting to net it. Once it detects the net, it has enough time to turn in the air and avoid being caught. Being a weak flier, however, means it cannot cover great distances for food, shelter or water. Another downfall of its feeble flight is that a swift breeze nearly halts its flight, and strong winds force it back to its roost.

DESCRIPTION: The Eastern Pipistrelle has a wingspan of only 19–22 cm. This bat is a yellow or drab brown colour overall. The hairs on its back are tricoloured, with dark bases, yellow brown in the middle and dusky grey tips. The wings, interfemoral membrane, ears, nose and feet are dark brown, but the leading edge of the wing is very light. It has a long and slender tragus and a keeled calcar. The contrast of its dark face and light fur gives the appearance that it is wearing a mask.

HABITAT: Eastern Pipistrelles are most common in shrubby areas and open woodlands close to waterbodies. Sometimes found close to cities, they are usually the first bat out in the evening and may even be seen in broad daylight. One great threat to these miniature bats is desiccation—they must always be

RANGE: These bats are found mainly from southern Ontario and Québec through all of the eastern states.

Total Length: 80–90 mm
Tail Length: 36–45 mm
Forearm: 31–35 mm
Weight: about 6 g

near water or they will die in less than a day.

FOOD: Eastern Pipistrelles feed on tiny flying insects such as flies, some beetles, leafhoppers and small moths. Because of the small size of this bat, large insects are inedible.

DEN: When roosting or hibernating, these bats can be found in caves, mines, crevices and old buildings. Some bats in northern areas migrate southward. Maternity colonies are found in crevices of rocky cliffs or in sheltered nooks of old buildings.

YOUNG: In June, females give birth to two young (rarely only one), usually in protected maternity colonies. These maternity colonies are not more than about 12 females, and sometimes a female may roost alone to bear her young. The young require their mother's care for several weeks until they are mature and ready for independence. Females who are lactating are at great risk of dehydration, and they can be seen any hour of the night at waterbodies near their roost.

SIMILAR SPECIES: The **Eastern Small-footed Bat** (p. 186) occupies similar habitats and may have similar coloration to the Eastern Pipistrelle, though it is larger and has a much longer tragus. The **Northern Bat** (p. 183) is slightly larger and has darker ears. The **Little Brown Bat** (p. 184) lacks the tricoloured hairs on the back.

DID YOU KNOW?

The erratic and jerky flights of the pipistrelle bats in Europe brought about the common name for bats as "flittermouse" or "fledermaus" in German.

INSECTIVORES & OPOSSUMS

This grouping, unlike the others in this book, actually encompasses two separate orders of mammals: shrews and moles belong to the order Insectivora (insectivores); opossums are in the order Didelphimorpha (New World opossums), which is part of the marsupials supergroup.

Opossum Family (Didelphidae)

The Virginia Opossum is the only marsupial in North America north of Mexico. Marsupials get their name from the marsupium, or pouch, in which newborns are typically carried. Because they usually have no placenta, marsupials bear extremely premature young that range from honeybee to bumblebee size at birth. Once in the marsupium, the young attach to a nipple and continue the rest of their development outside the uterus. There is also a difference in dentition between placental mammals and marsupials: early placental mammals typically have four premolars and three molars, while early marsupials have three premolars and four molars. Most of the world's marsupials live in Australia and New Zealand, with a few in South and Central America.

Virginia Opossum

Mole Family (Talpidae)

Moles are some of the most subterranean mammals. They spend the majority of their lives underground, and they have special adaptations that allow this to happen. Moles look a bit like large, rotund shrews, except their tails are proportionately shorter and their forelimbs are highly modified. The rotund appearance results from lack of a visible neck—the head merges smoothly with the body. This streamlined shape makes moving in underground tunnels much easier. The forelimbs have long claws and an enormous appearance; they are turned outward like paddles enabling moles to almost "swim" through soil. Their fur is lax, meaning they can move forward or backward in tunnels easily. Their eyesight is poor, as you might

Eastern Mole

guess from their tiny eyes, but their hearing is superb. Their most important sensory organ is the snout, which is flexible and usually hairless.

Moles consume up to twice their weight in food daily, and although earthworms are their primary food, they consume great numbers of pest insect larva.

Shrew Family (Soricidae)

Shrews first appeared in the times of the Cretaceous dinosaurs, and biologists consider the modern members to be most similar to the earliest placental mammals. Many people mistake shrews for very small mice. Shrews don't have a rodent's prominent incisors, however, and they generally have smaller ears and long, slender, pointed snouts.

Because they are so small, shrews lose heat rapidly to their surroundings, and their metabolisms surpass those of all other mammals. These tiny mammalian furnaces use energy at such a high rate that they may eat three times their own weight in invertebrate and vertebrate food each day. Some shrews have a neurotoxic venom in their saliva that enables them to subdue amphibians and mice that outweigh them. In turn, shrews are eaten by owls, hawks, foxes, Coyotes and weasels.

Shrews do not hibernate, but their periods of intense food-searching activity, which last 30 to perhaps 45 minutes, are interspersed with hour-long energy-conserving periods of deep sleep, during which the body temperature drops.

Of the shrews of Ontario, the Northern Short-tailed Shrew, Water Shrew and Arctic Shrew are reasonably easy to identify visually, provided you can get a long enough look at them. The other species must be distinguished from one another on the basis of tooth and skull characteristics, distribution and, to some extent, habitat, though in many cases the ranges overlap.

Water Shrew

Key to Shrews of Ontario

1a. Three unicuspid teeth visible in each side of upper jaw 2
1b. Four or five unicuspid teeth visible in each side of upper jaw 3
2a. Total length greater than 65 mm . . . . *Cryptotis parva* (p. 208)
2b. Total length less than 65 mm . . . . *Sorex hoyi* (p. 206)
3a. Tail less than ²/₅ of length of head and body . . . . *Blarina brevicauda* (p. 207)
3b. Tail more than ²/₅ of length of head and body 4
4a. Hind foot more than 18 mm, fur grayish, never truly brown . . . . . *S. palustris* (p. 202)
4b. Hind foot less than 18 mm, fur brownish 5
5a. Fur distinctly tricolored (back sharply darker than sides) . . . . *S. arcticus* (p. 205)
5b. Fur bicolored (back same as sides) 6
6a. Total length more than 110 mm . . . . *S. fumeus* (p. 204)
6b. Total length less than 110 mm . . . . *S. cinereus* (p. 201)

Virginia Opossum
Didelphis virginiana

Among the mammals of North America, the Virginia Opossum is unique because of its prehensile tail, maternal pouch, opposable "thumb" and habit of faking death. Famed by its portrayals in children's literature, the opossum is widely known but poorly understood. Few people realize that this animal is a marsupial, and that it is more closely related to the kangaroos and koalas of Australia than to any other mammal native to the U.S. or Canada.

Thanks to the many children's stories, we conjure up images of opossums hanging in trees by their tails. This behaviour is not nearly as common as the literature suggests. An opossum's tail is prehensile and strong, but it is unlikely to be used in such a manner unless the animal has slipped or is reaching for something.

The phrase "playing possum" is derived from the feigned death scene that is put on by a frightened opossum. If an opossum cannot scare away an intruder with fervent hissing and screeching, it will roll over, dangle its legs, close its eyes, loll its tongue and drool. Presumably, this death pose is so startling that the opossum will be left alone.

If you do much driving through opossum country, it should not be long before you encounter one. Unfortunately, opossums are frequent victims of roadway collisions. They are slow-moving animals that forage at night and find the bounty of road-killed insects and other animals hard to resist. With an abundance of food, opossums may become very fat. They draw upon their reserves in winter in colder parts of their range, but most of Ontario is still too cold for these creatures, with their naked ears and tails.

DESCRIPTION: The opossum is a cat-sized, grey mammal with a white face, long, pointed nose and long tail. Its ears are black, slightly rounded and nearly hairless. Its tail is rounded, scaly and prehensile. The legs, the base of the tail and patches around the eyes are black. Its overall appearance is grizzled from the mix of white, black and grey hairs.

HABITAT: Opossums favour moist woodlands or brushy areas near watercourses, but given a warm enough climate, opossums may be found almost anywhere, even in cities. Alternating warm and cold winters can cause an increase or decrease in their range here in Ontario.

FOOD: A full description of the opossum diet would include almost every-

RANGE: The opossum is found in southern Ontario and most of the eastern U.S. It was introduced in the western U.S. and now ranges along the entire West Coast as far north as B.C. and eastward along the Snake River into Idaho.

Total Length: 69–84 cm
Tail Length: 30–36 cm
Weight: 1.1–1.6 kg

thing organic. These omnivores eat invertebrates, insects, small mammals and birds, grain, berries and other fruits, grass and carrion.

DEN: By day, opossums hide in burrows dug by other mammals, in hollow trees or logs, under buildings or in rock piles. In colder parts of the region, they may remain holed up in a den for days during cold weather, but they do not hibernate.

YOUNG: Up to 25 young may be born in a litter after a gestation of 12 to 13 days. The young must crawl into the pouch and attach to one of the nipples (usually 13, but number may vary) if they are to survive. Only the young attached to a nipple can survive. After about three months in the pouch, an average of eight to nine young emerge, weighing about 160 g. Females mature sexually when they are six months to a year old.

SIMILAR SPECIES: No other mammal shares the combination of characteristics seen in the opossum. The young, newly emerged from the pouch, might be mistaken for rats, but rats do not have naked, black ears.

DID YOU KNOW?

At about the size of a honeybee at birth, an opossum begins life as one of the smallest baby mammals in North America.

Hairy-tailed Mole
Parascalops breweri

Total Length: 14–17 cm
Tail Length: 2.3–3.6 cm
Weight: 40–64 g

Hairy-tailed Moles are perfectly suited to life underground—understandably so, as they spend nearly all of their lives beneath the surface. Moles are as choosy about their earthen home as we are about what house we live in. Subsurface rocks do not bother Hairy-tailed Moles—they merely dig around them—but they are very fastidious about the soil type. They like sandy loam soils because the soil structure allows for sturdy burrows to be built. They avoid soils that have too much moisture or too much clay—burrows in this type of substrate are not durable and require too much maintenance.

DESCRIPTION: As its name suggests, this mole has a hairy tail. The tail is quite short and dark brown. The base of the snout and the feet are also dark brown. These brown areas may turn grey or white in older moles. The body is dark grey or nearly black above and only slightly paler below.

HABITAT: This mole favours well-drained soils in woodlands, brushy areas or meadows. It is sometimes found in golf courses and farmland.

FOOD: A voracious eater, the Hairy-tailed Mole consumes earthworms, grubs, insects and other invertebrates.

DEN: This mole lives in extensive tunnels underground. It is active all year, and in winter it retreats to deeper tunnels where the temperature is constant. A nest chamber is made for sleeping and giving birth.

YOUNG: Females produce one litter per year, usually in early summer. The young are altricial and require about one month to mature enough to leave the nest.

SIMILAR SPECIES: The **Eastern Mole** (p. 199) has a short, naked tail. The **Star-nosed Mole** (p. 200) has a long, furry tail and protrusions from the nose.

RANGE: The Hairy-tailed Mole is found in southern Ontario and Québec and from New England through the mountains to North Carolina and Tennessee.

Eastern Mole
Scalopus aquaticus

Total Length: 10–20 cm

Tail Length: 2–3 cm

Weight: 50–120 g

The Eastern Mole is the most widespread of any mole species in North America, hence its alternate name, the Common Mole. Unfortunately, its Latin name is not as accurate as its two common names. The species name, *aquaticus*, was erroneously applied by the first people to describe this mole. One of the earliest specimens was found drowned in a well, and close inspection of its forefeet revealed partial webbing between the toes. This evidence led people to think the mole was aquatic. Logically, of course, the webbing would help with digging through soil, and the specimen's drowning should have proved that it was not aquatic.

ALSO CALLED: Common Mole.

DESCRIPTION: This rotund mole is quite variable in size, but its short, nearly naked tail helps distinguish it from the other two moles in Ontario. The colour of its fur varies regionally. Here in Ontario, its fur is predominantly grey in colour; moving south, the fur colour becomes brown or tan. Its front feet are broader than they are long. Its nose is long, flexible and nearly naked.

HABITAT: This mole usually prefers open areas such as meadows and sparse woodlands. It favours loose and well-drained soils.

FOOD: This mole forms extensive burrows for feeding, and its primary food is earthworms, though it also consumes insects, slugs and other invertebrates.

DEN: The Eastern Mole makes a shallow underground nest, usually protected by a surface boulder or log. Some nests and burrows are used repeatedly, but others may be abandoned after one use.

YOUNG: Mating occurs in February or March, and after a gestation of 30 to 42 days the female has a litter of two to five young. The young are altricial, but they develop rapidly. By one month, the young are on their own.

SIMILAR SPECIES: The **Hairy-tailed Mole** (p. 198) and the **Star-nosed Mole** (p. 200) both have longer furred tails.

RANGE: The Eastern Mole is found in much of the eastern U.S. and just into Ontario at the southernmost border (in Essex county).

Star-nosed Mole
Condylura cristata

Total Length: 15–21 cm
Tail Length: 5.3–8.4 cm
Weight: 30–75 g

The Star-nosed Mole is a fine example of the incredible variation and diversity that exists in the animal kingdom. Unlike all other moles, the Star-nosed Mole has developed a ring of "feelers" around its nose. At first glance, its nose looks gruesome and medusa-like, but in reality, this interesting structure is an elegantly designed supersensor. Each fleshy appendage can be collapsed or extended individually, unaffected by the movement of the ones next to it. They move continuously in all directions to acutely and perfectly interpret the mole's surroundings.

Star-nosed Moles hunt for food either underground or in water, and their noses are perfectly suited for either habitat.

DESCRIPTION: There is no mistaking this distinctive mole. The 22 tentacle-like projections in a ring around the nose are pink and fleshy. Its fur is silky black, and its tail is long and hairy. Its front feet are powerful and have long claws.

HABITAT: This mole usually prefers wet areas such as marshes, low fields or humid woodlands. Sometimes it inhabits drier areas.

FOOD: The primary food of these moles is earthworms, though they also consume insects, slugs and other invertebrates, including some aquatic invertebrates.

DEN: The Star-nosed Mole forms extensive burrows for feeding, but its leaf and grass nests are usually in a clump of vegetation or hummock on the surface of the ground.

YOUNG: Females have one litter per year of four to seven young, usually in early spring. Gestation is about 45 days. The young are altricial, but they develop rapidly. By the third week, the young have left the nest.

SIMILAR SPECIES: Both the **Hairy-tailed Mole** (p. 198) and the **Eastern Mole** (p. 199) lack the distinctive nose and long, well-haired tail.

RANGE: Star-nosed Moles are found in southeastern Canada and the northeastern U.S. Southerly populations are found in the Appalachians and on the Georgia coast.

Masked Shrew

Sorex cinereus

Total Length: 7–11 cm
Tail Length: 2.5–5.1 cm
Weight: 2–7 g

The Masked Shrew may be the most common shrew throughout Ontario. In spite of its abundance in this province, you are most likely to see one dead in spring; starvation in late winter claims many, and their tiny bodies lay waiting to be recycled in the renewal of spring. To balance the high mortality rates in late winter and the high year-round predation, these shrews have high fecundity. They mate from May to October, and the ones that are born in autumn have a good chance of making it to spring when they too will mate.

ALSO CALLED: Cinereus Shrew, Common Shrew.

DESCRIPTION: These medium-sized shrews have dark brown backs, lighter brown sides and slightly lighter underparts. The winter coat is paler, and the fur is short and velvety. It has a long, flexible snout, tiny eyes, small feet and a bicoloured tail, which is dark above and light below. A few may have a dark patch on the nose—the "mask" for which the shrew is named.

HABITAT: The Masked Shrew favours forests, either coniferous or deciduous, and sometimes tallgrass plains or brushy coulees.

FOOD: Insects account for the bulk of the diet, but this shrew also eats significant numbers of slugs, snails, young mice, carrion and even some vegetation.

DEN: The nest, located under logs, in debris, between rocks or in burrows, is about 5–10 cm in diameter and looks like a woven grass ball. The nest does not have a central cavity; the shrew simply burrows to the inside.

YOUNG: Mating occurs from April to October, and, with a gestation of about 28 days, a female may have two or three litters a year. The four to eight young are born naked, toothless and blind. Their growth is rapid: eyes and ears open in just over two weeks, and the young are weaned by three weeks.

SIMILAR SPECIES: Most shrews look very similar. Without a specimen and a technical key, it is almost impossible to identify a shrew reliably.

RANGE: The Masked Shrew is found across most of Alaska and Canada. Its range extends south into northern Washington, through the Rockies and across most of the northeastern U.S.

Water Shrew

Sorex palustris

As everyone would agree, most of the shrews in Ontario have few distinguishing characteristics. Water Shrews, however, are an exception in the region's shrewdom—these finger-sized heavyweights are so unusual in their habits that they deserve celebrity status.

While other shrews prefer to wreak terror exclusively on the small vertebrates and invertebrates roaming on land, Water Shrews also take the plunge to feed upon aquatic prey. The Water Shrew is a particularly fierce predator, ably seizing not only insect nymphs, but even sticklebacks and other small fish. The shrew drags the catch onto land, where it is quickly consumed. The Water Shrew is specially adapted for swimming: small hairs on the hindfeet widen the foot and create a flipper effect for propulsion. The shrew is very powerful and can easily out-swim most prey species. Once it is out of the water, this shrew's fringed feet serve as a comb with which to brush water droplets out of the fur.

An easy shrew to recognize, the Water Shrew can be seen beneath overhangs along flowing waters, particularly small creeks and backwaters. If you are walking along these shorelines, you might see a small, black bundle rocket from beneath the overhang into the water. The motion at first suggests a frog, but the Water Shrew tends to enter the water with more finesse, hardly producing a splash. Often, the shrew first runs a short distance across the surface of the water before diving in. Some voles and mice are also scared into or across water in this way, but even at a quick glance you can distinguish this shrew from those rodents by its velvety black colour.

DESCRIPTION: The Water Shrew is one of the largest shrews in Ontario. It has a velvety, black back and contrasting light brown or silver underparts. The third and fourth toes of the hindfeet are slightly webbed, and a stiff fringe of hairs around the hindfeet aid in swimming. Males tend to be somewhat larger than females.

HABITAT: This shrew can be found alongside flowing streams with undercut, root-entwined banks, in sphagnum moss on the shores of lakes and occasionally in nearly dry streambeds or tundra regions.

FOOD: Aquatic insects, spiders, snails, other invertebrates and small fish form the bulk of the diet. With true shrew

RANGE: This transcontinental shrew ranges from southern Alaska to Labrador and south along the Cascades and Sierra Nevada to California, along the Rocky Mountains to New Mexico and along the Appalachians almost to Georgia.

Total Length: 14–17 cm
Tail Length: 6–9 cm
Weight: 9–19 g

frenzy, this scrappy water lover may even attack fish half as large as itself.

DEN: This shrew dens in a shallow burrow in root-entwined banks, in sphagnum moss shorelines or even in the woody debris of Beaver lodges (see p. 154). The nest is a spherical mound about 10 cm in diameter and composed of dry vegetation, such as twigs, leaves and sedges.

YOUNG: The Water Shrew breeds from February until late summer. The females have multiple litters each year. Females born early in the year usually have their first litter in that same year. Litters vary in size from five to eight

young, and, as with other shrews, the young grow rapidly and are on their own in a few weeks.

SIMILAR SPECIES: All other shrews in the region lack the velvety, black fur of the Water Shrew. The **Arctic Shrew** (p. 205) may have deep brown or black fur, but only on the top of its back.

DID YOU KNOW?

Both terrestrial and aquatic animals prey on Water Shrews. Weasels, Minks and otters can catch them, as can large trout, bass, walleye and pike.

Smoky Shrew
Sorex fumeus

Total Length: 11–13 cm
Tail Length: 3.7–5.2 cm
Weight: 6.5–9.9 g

These frenetic shrews seem to have inexhaustible energy and audacity. Win or lose, they are known to persistently attack almost any small animal they encounter. Large salamanders—significantly larger than this shrew—are killed with a crushing bite to the spinal cord at the neck.

The Smoky Shrew, despite its daring, meets its match with large *Peromyscus* mice (pp. 142–44). Try as it might, this shrew seems unable to beat a large mouse, and it eventually turns away defeated. This shrew is quite talkative, and when it is alarmed it utters a high-pitched, grating note. Even while it forages, it continually twitters, although the noise is almost inaudible to humans.

DESCRIPTION: Unlike the Masked Shrew (p. 201), this shrew's colour is nearly uniformly brown or greyish. In summer the fur is browner, and in winter the fur is greyer. It has a long tail that is brown on top and tawny or sometimes yellowish below.

HABITAT: The Smoky Shrew can be found in a variety of different moist conditions, such as marshes, humid wooded areas and along streams.

FOOD: The majority of the Smoky Shrew's diet is soft-bodied, tender invertebrates, such as insects and their larva, earthworms and slugs.

DEN: In a safe place, such as in a hollow log or under rocks, this shrew makes a small, leafy nest. It also makes burrows with entrances about the size of a dime.

YOUNG: Mating occurs in spring, especially in March. About five or six young are born in the summer months after a gestation of 20 days. The young grow quickly, and although they are sexually mature in their first year, females do not bear young until they are two.

RANGE: Smoky Shrews are found in central and southern Ontario and Québec, New Brunswick, Nova Scotia, New England and through the Appalachians to northeast Georgia.

SIMILAR SPECIES: All other shrews are difficult to differentiate—the best clue is this shrew's nearly uniform coloration. The **Northern Short-tailed Shrew** (p. 207) is larger and greyer.

Arctic Shrew

Sorex arcticus

Total Length: 10–12 cm
Tail Length: 3.8–4.5 cm
Weight: 5–14 g

Arctic Shrews may be the most handsome of all North American shrews. If you were able to observe one of these animals, you could even tell the season by the shrew's colour. Although many weasels and hares change colour seasonally, it is quite unusual for a shrew. Not only is the Arctic Shrew's winter coat longer and denser than its summer coat, it is also more vibrant, with a black back, brown sides and a white or greyish belly. The full summer coat is less striking, with a brown back and grey underparts.

ALSO CALLED: Saddle-backed Shrew.

DESCRIPTION: The tricoloured body of this stocky shrew makes it one of the easiest shrews to recognize: the back is chocolate brown in summer and glossy black in winter; the sides are grey brown year-round; the undersides are ashy grey in summer and silver white in winter. The tail is cinnamon coloured year-round. Females are usually slightly larger than males.

HABITAT: This shrew typically inhabits moist areas of the boreal forest or its edges. Outside forested regions, the Arctic Shrew takes to open areas, dried-out sloughs and streamside habitats among shrubs.

FOOD: The Arctic Shrew feeds primarily on invertebrates, such as insects, snails, slugs and even some carrion.

DEN: The spherical, grassy nest, 6–10 cm in diameter, is built in a small pocket in or under logs, under debris or in rock crevices.

YOUNG: Breeding takes place between May and August, and females generally have two litters of 4 to 10 young in a season. Females born early in the year may have their first litter in late summer of that same year, but most females do not breed until the next year.

SIMILAR SPECIES: The tricoloured pelage of the Arctic Shrew—a dark back, lighter sides and a still lighter belly—best distinguishes it from other shrews. It also tends to be heavier and stockier than most other shrews. The **Water Shrew** (p. 202) is velvety black over its back.

RANGE: The Arctic Shrew is found from the southeastern Yukon across central Canada to Newfoundland and south to Minnesota, Wisconsin and parts of North Dakota and Michigan.

Pygmy Shrew
Sorex hoyi

Total Length: 5–6 cm
Tail Length: 2.5–3.2 cm
Weight: 2–7 g

Weighing no more than a penny, the Pygmy Shrew represents the furthest degree of miniaturization in mammals. It is considered to be the smallest of all North American mammals. The Dwarf Shrew (*S. nanus*), found in southern areas of the U.S., may weigh less, but it is longer than the Pygmy Shrew.

In spite of its size, the Pygmy Shrew is every bit as voracious as other shrews; one female on record ate about three times her body weight each day for 10 days. The Pygmy Shrew may also be one of the rarest shrews in North America.

DESCRIPTION: This tiny shrew is primarily reddish to greyish brown. The colour grades from darkest on the back to somewhat lighter underneath. It is usually greyer in winter. The third and the fifth unicuspid teeth are so reduced in size that they may go unnoticed.

RANGE: The Pygmy Shrew occurs from Alaska east to Newfoundland and south to Colorado, the Appalachians and New England.

HABITAT: The Pygmy Shrew lives in a variety of different habitats, moist to dry and forested to open, including deep spruce woods, sphagnum bogs, grassy or brushy areas, cattails and rocky slopes.

FOOD: This shrew feeds primarily on both larval and adult insects, but earthworms, snails, slugs and carrion often make up a significant portion of the diet.

DEN: The spherical, grassy nest, 6–10 cm in diameter, may be under logs, under debris or in rock crevices. Unlike the nests of many other mammals, there is no rounded cavity inside this grassy ball; instead the shrew simply burrows its way in among the grass.

YOUNG: Breeding takes place from May until August, and 4 to 10 young are born in June, July or August. Females generally have only one litter a year. Young born early in the year may have a late-summer litter, but most females do not mate until the following year.

SIMILAR SPECIES: Other small shrews may be impossible to distinguish unless measurements are obtained.

Northern Short-tailed Shrew
Blarina brevicauda

Total Length: 9.6–14 cm

Tail Length: 2–3 cm

Weight: 14–29 g

When a Northern Short-tailed Shrew digs its burrow, it digs rapidly with its front feet in a blur of motion. As the dirt accumulates underneath itself, the shrew kicks it out of the entranceway with its hindfeet. As it tunnels deeper and more dirt accumulates, the only way it can get rid of the loose dirt is to turn a sideways somersault in the burrow and bulldoze the earth out to the ground surface using its forehead.

The Northern Short-tailed Shrew has a very high metabolism—as all shrews do—and the rapid activity tires it quickly. Throughout the process of digging a burrow, the shrew must take frequent breaks to have a short nap.

DESCRIPTION: This shrew is one of the largest in North America. Its robust body is uniformly lead grey in colour. It has a noticeably short tail that is similarly coloured.

HABITAT: This shrew inhabits many regions, such as woodlands, fields and marshy areas. In hot climates, it remains in moist areas.

FOOD: This voracious and even vicious shrew eats not only the standard shrew fare of insects and soft invertebrates, but it is known to attack young rabbits and once even a 38 cm-long garter snake!

DEN: In a rock crevice or hollow log, this shrew builds a bulky leaf or grass nest. It also digs burrows and runways, which it patrols in search of food.

YOUNG: Mating may occur throughout the year, but the peak is in spring. Litter size is four to eight young, and gestation varies from 17 to 21 days. Young are about the size of honeybees at birth, and they are nearly mature at 50 days old.

SIMILAR SPECIES: This shrew is distinctive because of its large size and short tail. Most other shrews are smaller, and the **Smoky Shrew** (p. 204) is browner.

RANGE: This shrew is found in southeastern Canada, from Saskatchewan to the coast and south to Nebraska, Kentucky, the Appalachians to Alabama and the upper mid-Atlantic.

Least Shrew
Cryptotis parva

Total Length: 6.9–8.9 cm
Tail Length: 1.9–2.2 cm
Weight: 4–6.5 g

New to most people in Ontario, the Least Shrew is only known from a couple of records in the southern part of the province. The Least Shrew is much better known in the eastern states, where it can be found in a wide variety of habitats. Unfortunately, this shrew is not well-studied, and information about it is limited. The information below has been taken from populations south of Ontario.

The Least Shrew is the only member of the genus *Cryptotis*, and it has an extremely short tail when compared to most other shrews.

DESCRIPTION: The Least Shrew is very small with a brownish-grey back that grades to lighter colours on the sides and underside. In summer this shrew is browner; in winter it is greyer. It has a short tail that is less than half the head and body length. If you raise the upper lip on the side of the snout, three unicuspid teeth are visible. The fourth is present as well, but it is hidden behind the third.

HABITAT: This tiny shrew seems to prefer open deciduous forests, wooded ravines and some grassy or wet areas.

FOOD: The Least Shrew feeds mainly on invertebrates, such as insects of all types, spiders, millipedes, worms, snails and slugs.

DEN: The den is often found in soft soil, among rocks or under woody debris. The nest chamber is exceedingly small, and the entrance to the burrow is small and indistinct. There are some records of several Least Shrews building and sharing the same burrow.

YOUNG: Little is known about this shrew's reproduction, but it is probably similar to other shrews. Mating occurs from March through November, with females having multiple litters a year.

SIMILAR SPECIES: Most shrews are difficult to distinguish without a specimen in hand and a technical key (p. 195) for reference.

RANGE: Least Shrews are found in most of the eastern U.S., and records exist from Long Point in Ontario.

Glossary

ARBOREAL: living in or pertaining to trees.

BUFF: a dull, brownish yellow.

CACHE: a place in which food is hidden for future use; food hidden in such a place.

CALCAR: in bats, a small projection from the inner side of each hind foot into the membrane between the hind legs.

CANID: a member of the dog family (Canidae).

CARNIVOROUS: flesh-eating.

CERVID: a hoofed mammal of the deer family (Cervidae).

COLONY: a group of animals living together and interacting socially.

CONIFEROUS: pertaining to needle-leaved, cone-bearing trees (e.g., fir, spruce, pine).

DECIDUOUS: pertaining to trees that shed their leaves in autumn (e.g., oak, maple, elm).

DORMANCY: a state of inactivity, with greatly slowed metabolism, respiration and heart rate.

DORSAL: pertaining to the back or spine (compare *ventral*).

DREY: a spherical tree nest made of leaves, twigs and moss.

ECHOLOCATION: the ability of some animals (including bats and cetaceans among mammals) to detect an object by emitting sound waves and interpreting the returning echoes, which are changed from bouncing off the object.

ENDANGERED: said of a species or subspecies that is facing imminent extirpation or extinction.

EXTINCT: said of a species that no longer exists anywhere.

EXTIRPATED: said of a species that no longer exists in a given geographic area but still survives elsewhere in the world.

GESTATION: the time of pregnancy, from conception to birth.

GREGARIOUS: preferring to living in large groups with other individuals of the same species; sociable.

GRIZZLED: said of mostly dark fur that is sprinkled or streaked with gray or another light colour.

GUARD HAIRS: long, coarse hairs that help protect a mammal's underfur from the weather.

HABITAT: the environment in which an animal or plant lives.

HERBACEOUS: pertaining to plants that lack woody stems.

HERBIVOROUS: plant-eating.

HIBERNACULUM: the den in which an animal hibernates.

HIBERNATION: winter dormancy.

HIERARCHY: a social order; the ranking of individuals by social status.

HOME RANGE: the total area through which an individual animal moves during its usual activities (compare *territory*).

INTERBREED: for individuals of different species to mate with each other.

MEMBRANE: a thin, flexible layer, such as the skin of a bat's wings.

MIGRATION: the journey that an animal undergoes to get from one region to another, usually in response to seasonal and reproductive cycles.

NOCTURNAL: active at night.

OMNIVOROUS: feeding on both plant and animal material.

PALMATE: branching like the fingers of a human hand.

PELAGE: the fur or hair of a mammal.

PINNIPED: a member of the subgrouping of the order Carnivora that encompasses all seals, sea-lions and walruses.

PREDATOR: an animal that kills its prey (compare *scavenger*).

RUNWAY: a beaten path made by the repeated travels of small animals.

SCAT: a fecal pellet or dropping; feces.

SCAVENGER: an animal that feeds on animals it did not kill (compare *predator*).

SUBNIVEAN: under the snow (but above the ground).

SUBTERRANEAN: underground.

TERRITORY: a defended area within an animal's home range.

TRAGUS: a lobe projecting upward from inside the base of the ears, as in bats.

UNDERFUR: a thick, insulating undercoat of fur.

UNGULATE: a hoofed mammal.

UNICUSPID: in shrews, any of the small teeth between the two front teeth and the large rear teeth.

UROPATAGIUM: the fold of skin that stretches from a bat's hind legs to its tail.

VENTRAL: pertaining to the belly (compare *dorsal*).

Index of Scientific Names

Page numbers in **boldface** type refer to the primary, illustrated species accounts.

Index of Common Names

Page numbers in **boldface** type refer to the primary, illustrated species accounts.